Employability for
PhD Students in STEM

Online at: https://doi.org/10.1088/978-0-7503-5542-1

Employability for PhD Students in STEM

Jatinder Vir Yakhmi
Formerly at Bhabha Atomic Research Centre (BARC), Mumbai 400 085, India

IOP Publishing, Bristol, UK

© IOP Publishing Ltd 2024. All rights, including for text and data mining (TDM), artificial intelligence (AI) training, and similar technologies, are reserved.

This book is available under the terms of the IOP-Standard Books License

No part of this publication may be reproduced, stored in a retrieval system, subjected to any form of TDM or used for the training of any AI systems or similar technologies, or transmitted in any form or by any means, electronic, mechanical, photocopying, recording or otherwise, without the prior permission of the publisher, or as expressly permitted by law or under terms agreed with the appropriate rights organization. Certain types of copying may be permitted in accordance with the terms of licences issued by the Copyright Licensing Agency, the Copyright Clearance Centre and other reproduction rights organizations.

Permission to make use of IOP Publishing content other than as set out above may be sought at permissions@ioppublishing.org.

Jatinder Vir Yakhmi has asserted his right to be identified as the author of this work in accordance with sections 77 and 78 of the Copyright, Designs and Patents Act 1988.

ISBN 978-0-7503-5542-1 (ebook)
ISBN 978-0-7503-5540-7 (print)
ISBN 978-0-7503-5543-8 (myPrint)
ISBN 978-0-7503-5541-4 (mobi)

DOI 10.1088/978-0-7503-5542-1

Version: 20241101

IOP ebooks

British Library Cataloguing-in-Publication Data: A catalogue record for this book is available from the British Library.

Published by IOP Publishing, wholly owned by The Institute of Physics, London

IOP Publishing, No.2 The Distillery, Glassfields, Avon Street, Bristol, BS2 0GR, UK

US Office: IOP Publishing, Inc., 190 North Independence Mall West, Suite 601, Philadelphia, PA 19106, USA

Dedicated to my wife, Mrs Amar Upasana Yakhmi (1948–2022)

Upasana would have been thrilled to see this book in its final shape! I started writing it while she was still around. I dedicate this book to her, recognizing her unfailing support to me, all through our wedded life of over 50 years, during which she encouraged me to do my science, or write articles/books. As a spouse, she stood behind me like a rock during difficult phases of my life. A graduate in Home Science, she was an expert home-maker. Upasana was a caring mother; our children – son Ashish and daughter Geetika knew she was there for them, whenever they needed her help.

IOP Publishing's ebooks and journals follow the C4DISC Guidelines on Inclusive Language and Images in Scholarly Communication. We no longer use gendered pronouns such as he/him/his, she/her/hers, himself/herself, and instead use they/their/them/theirs/themselves, all of which are gender neutral and can be correctly used in a singular form in standard English grammar.

Contents

Preface	xvi
Author biography	xix

1 Learning science: a passion for STEM subjects: why do a PhD? 1-1

1.1	Scientific literacy	1-1
1.2	Passion for science and STEM subjects	1-2
1.3	Benefits of science education	1-2
1.4	STEM education	1-3
1.5	STEM is the future	1-3
1.6	Why do a PhD?	1-4
1.7	Catch them young for STEM	1-5
1.8	A PhD degree is a well-trodden path for employment	1-6
1.9	PhD makes a difference	1-6
1.10	Passion for STEM—the lure of a PhD in STEM	1-7
1.11	Why do a STEM PhD?	1-7
1.12	Employment after a STEM PhD	1-8
1.13	Conclusion	1-9
	References	1-9

2 Sailing through PhD work successfully 2-1

2.1	Introduction: doing a PhD in STEM for entry into exciting careers		2-1
2.2	Essential requirements for smooth sailing during doctoral work		2-3
	2.2.1	Choosing a strong problem to do research towards a PhD thesis	2-3
	2.2.2	Settle for a lab that has requisite facilities to conduct PhD work	2-3
	2.2.3	Try working with a reputed PhD guide/supervisor	2-4
	2.2.4	Avoiding some pitfalls while choosing a supervisor	2-4
	2.2.5	Collect a lot of original data for authenticity of results and reliable conclusions	2-5
	2.2.6	Publish as many journal publications as possible based on thesis work, of these, the majority should have the PhD candidate as first co-author	2-5
2.3	Other relevant points to be followed for unhindered PhD work		2-6
	2.3.1	Acquiring of skills is important for doctoral candidates	2-6
	2.3.2	STEM PhD does provide an edge	2-7

	2.3.3	Learn to share facilities and consumables in the lab	2-7
	2.3.4	Be humble during field work	2-7
	2.3.5	Need for a mentor	2-8
	2.3.6	Problems with mentorship	2-9
	2.3.7	Attending conferences and networking boosts employability of a STEM PhD	2-10
	2.3.8	Managing time is crucial for PhD students	2-10
	2.3.9	Feeling lonely and depressed during PhD	2-11
	2.3.10	Be clear and selective while making commitments at the workplace	2-14
2.4	A PhD degree should lead to careers and jobs		2-14
	2.4.1	Careers in education research	2-15
	References		2-15

3 Using PhD final year and just after to master useful skills—writing and public speaking — 3-1

3.1 Role of effective communication in science is crucial for a STEM PhD — 3-1
3a.1 Learning to write on research topics: avoiding plagiarism — 3-2
   3a.1.1 Academic integrity — 3-3
   3a.1.2 Avoiding plagiarism — 3-3
   3a.1.3 Copyrights — 3-4
3a.2 Writing of manuscripts to publish in a journal — 3-5
   3a.2.1 Peer review — 3-8
   3a.2.2 Committing fraud, paper mills, and the after-effects of retractions — 3-9
      3a.2.2.1 Predatory journals and zombie papers — 3-9
      3a.2.2.2 Dual publication fraud — 3-10
      3a.2.2.3 Retractions — 3-10
   3a.2.3 Open-access publishing — 3-10
   3a.2.4 Publishing a monograph — 3-11
3a.3 Writing of a CV and a résumé — 3-11
   3a.3.1 A curriculum vitae — 3-11
   3a.3.2 Résumés — 3-12
3a.4 Writing of thesis/dissertation — 3-14
   3a.4.1 Why writing of a dissertation becomes so tough for some candidates — 3-14
   3a.4.2 Adapting a dissertation for publication, and papers arising from PhD work — 3-16

3a.5	Composing an application for a research grant	3-16
	3a.5.1 Project proposal	3-17
3a.6	Writing applications for jobs—a cover letter	3-17
3a.7	The letters of recommendation	3-18
3a.8	Research statement	3-18
3a.9	Preparing a career portfolio	3-19
3a.10	Becoming a PI	3-20
3b.1	Public speaking with confidence	3-20
	3b.1.1 Public speaking skills for doctoral students in STEM	3-20
	3b.1.2 Useful hints for successful public speaking	3-22
	3b.1.3 Fear of speaking in public	3-23
3b.2	Making a conference presentation—oral or poster	3-24
	3b.2.1 Oral presentation	3-25
	3b.2.2 Poster presentation	3-26
3b.3	Making full use of participation in a conference: learning to network	3-27
3b.4	Preparedness for a job in academia/research lab or in industry	3-28
	3b.4.1 Industry job interview	3-29
	3b.4.2 Facing an academic job interview, online	3-30
3b.5	Handling tricky questions during a job interview	3-30
3b.6	Transferable skills for industry jobs	3-32
	3b.6.1 Teaching skills	3-32
	3b.6.2 Skills from STEM education	3-33
	3b.6.3 Skills in science of communication	3-33
3b.7	Career guidance for STEM PhDs	3-34
	References	3-35

4	**Acquiring multidisciplinary skills to beat the 'single-subject cocoon' trap and ability to stay ahead of exponential technologies**	**4-1**
4.1	Introduction	4-1
	4.1.1 Multidisciplinary eco-system for jobs	4-1
	4.1.2 AI, ML, and ChatGPT	4-3
	4.1.3 Robots	4-7
	4.1.4 5G technology	4-8
	4.1.5 Resource-guzzler AI	4-9
4.2	Influence of AI on jobs	4-9
	4.2.1 AI engineers	4-10

		4.2.2	Will AI replace humans or complement humans?	4-11
		4.2.3	Which human skills AI is unlikely to master?	4-13
		4.2.4	With some efforts STEM PhDs can remain unbeaten and stay ahead of AI	4-14

4.3	Beating the 'single-subject cocoon' trap for enhanced employment opportunities		4-14
	4.3.1	Role of multidisciplinary expertise in job opportunities, typically for STEM PhDs	4-15
	4.3.2	Electric vehicles and batteries—opportunity for jobs for STEM PhDs	4-16
		4.3.2.1 Batteries and lithium	4-18
4.4	For continued employability STEM PhDs should upskill their job skills regularly		4-19
4.5	New jobs to be created by AI		4-20
4.6	Typical examples of new emerging job opportunities for STEM PhDs		4-21
	4.6.1	AI and ML in agriculture	4-21
	4.6.2	AI in radiology	4-21
	4.6.3	Using machine learning to identify new deposits of in-demand minerals	4-21
	4.6.4	Thin films for computer chips	4-22
	4.6.5	Designing medicines that can go across the blood–brain barrier	4-22
	4.6.6	Neuromusculoskeletal prosthesis	4-22
	4.6.7	Programmable bacteria to kill cancerous tissue	4-22
	4.6.8	Quantum technology	4-23
	4.6.9	Biocomputers	4-23
	4.6.10	Mapping of icebergs	4-24
	4.6.11	Jobs at LHC and at fusion tokamaks like ITER	4-24
	4.6.12	EVs	4-24
4.7	Societal benefits: improving the quality of life for masses		4-25
	4.7.1	Climate equity specialist—a job	4-25
	4.7.2	Teaching science at school	4-25
4.8	Conclusions		4-26
	References		4-26

5	**Mentoring, innovations, patents, entrepreneurships, and jobs**		**5-1**
5.1	Introduction		5-2
5.2	What is an innovation?		5-3
	5.2.1	Innovations can be risk-prone	5-4
	5.2.2	Encouraging innovations	5-5

5.3	Patents		5-5
	5.3.1	IP rules, copyrights, and patents	5-7
5.4	Prototypes		5-7
5.5	Entrepreneurship		5-7
	5.5.1	Mentoring for entrepreneurship	5-8
	5.5.2	How do entrepreneurs succeed in their pitches?	5-8
	5.5.3	Spreading entrepreneurial literacy	5-9
5.6	Ideas for entrepreneurships and start-ups by STEM PhDs		5-9
5.7	Jobs		5-11
	5.7.1	Career guidance	5-11
	5.7.2	Jobs for STEM PhDs	5-11
		5.7.2.1 Postdoc	5-11
		5.7.2.2 Successful transition from postdoctoral position into a faculty job	5-12
		5.7.2.3 Bridging the gap between academic research and industry	5-13
		5.7.2.4 Interface of academe and industry	5-14
		5.7.2.5 Postodoc versus industry jobs	5-15
5.8	Innovations by STEM PhDs towards SDGs for techno-entrepreneurships and jobs		5-16
	5.8.1	Health sector and health-care	5-17
	5.8.2	Agricultural sector	5-19
	5.8.3	Chips	5-20
	5.8.4	Biology for chemical engineers and sustainability	5-20
	5.8.5	Dementia villages	5-21
	5.8.6	Gene-editing	5-21
	5.8.7	Biotech with AI	5-21
5.9	New opportunities in STEM jobs by solving existential challenges		5-22
	5.9.1	Climate	5-22
		5.9.1.1 Wildfires	5-23
		5.9.1.2 Droughts	5-24
		5.9.1.3 Heatwaves	5-24
		5.9.1.4 Sea ice melting	5-24
		5.9.1.5 Deep sea mining	5-24
		5.9.1.6 Air turbulence	5-25
	5.9.2	Looming AI	5-25
	5.9.3	Smartphone-distracted driving	5-25

5.10	Opportunities for new jobs in the electrical vehicles sector	5-26
	5.10.1 Jobs being created by EVs in other sectors	5-26
	5.10.2 Jobs from tackling the devil in the EVs	5-27
	5.10.3 Scope for R&D in the EV sector	5-27
5.11	The new scenario under the emergence of AI and ChatGPT	5-28
5.12	A case study of Unilever	5-28
5.13	Job security	5-30
	References	5-31

6 Models adopted to upend education for supporting growth of graduate jobs — 6-1

6.1	Introduction	6-1
6.2	Jobs crisis post-PhD, and how to fix it	6-2
	6.2.1 Multi-disciplinary education	6-2
	6.2.2 Strengthening the skills of writing and drafting	6-2
	6.2.3 Mathematics is essential for employability	6-3
	6.2.4 Financial literacy	6-3
6.3	Country-wide efforts	6-4
	6.3.1 France	6-4
	6.3.2 India	6-4
	6.3.3 USA	6-5
	6.3.4 Germany	6-6
	6.3.5 China	6-6
	6.3.6 UK	6-6
	6.3.7 Malaysia	6-7
	6.3.8 Vietnam	6-7
6.4	Enhancing employability of STEM graduates and PhDs	6-7
	6.4.1 By staying on top of technology (example—electronic design engineers)	6-7
	6.4.2 By coaching them young	6-8
	6.4.3 Through institutional efforts	6-9
	6.4.3.1 To make cities healthier	6-9
	6.4.3.2 To study global warming	6-9
	6.4.3.3 Protection of women's health	6-9
	6.4.4 By acquiring new skills for STEM PhDs for jobs	6-10
	6.4.5 Fusion	6-11
	6.4.6 By acquiring additional skills through attending new courses	6-11

6.5	New doors/opportunities for STEM graduates, including those with disabilities			6-12
6.6	Skills for employability in industry			6-13
6.7	Apprenticeships and training in industry, for facilitating employment			6-14
	References			6-15

7 Overcoming disruptions against career growth — 7-1

7.1	Employment and career options for STEM PhDs			7-1
	7.1.1	Jobs in industry		7-2
	7.1.2	Transferable skills of value in business companies		7-3
		7.1.2.1	STEM PhDs do have transferable skills useful to the business world	7-3
7.2	Disruption due to the Covid-19 pandemic, or other disasters: disaster-proofing			7-4
	7.2.1	Covid-19 and workplace changes		7-5
	7.2.2	Mothers affected most among staff during Covid-19		7-5
	7.2.3	Economic anxiety in the wake of Covid-19, and non-lab positions		7-6
	7.2.4	Wars		7-6
	7.2.5	Disaster-proofing		7-6
7.3	Lay-offs, quiet hiring			7-6
	7.3.1	Lay-offs		7-6
		7.3.1.1	Advice to individual job seekers in the face of lay-offs	7-7
	7.3.2	Quiet hiring		7-8
7.4	Disabled			7-9
	7.4.1	Hurdles galore for the disabled		7-9
	7.4.2	Innovative new solutions for disabled		7-9
7.5	Debt ceiling			7-10
7.6	Bias towards gender, black, Asian, candidates coming under adult education			7-11
	7.6.1	Gender bias		7-11
		7.6.1.1	General	7-11
			7.6.1.1.1 Women are overlooked in STEM	7-11
			7.6.1.1.2 Women with PhD in STEM	7-12
			7.6.1.1.3 Bias against middle-aged women	7-12
			7.6.1.1.4 Patents by women	7-13
			7.6.1.1.5 How to rectify the bias against gender	7-13
		7.6.1.2	Mothers in science	7-13

	7.6.1.3	Women in STEM in Australia	7-13
	7.6.1.4	STEM recharge project: UK women can return to STEM careers	7-14
7.6.2	Bias towards black and Asian workers		7-14
7.6.3	Adult education		7-14

7.7 Quiet quitting, working from home, job-switching, great resignation ... 7-15
    7.7.1 Quiet quitting ... 7-15
    7.7.2 Job hopping ... 7-16
    7.7.3 Great resignation ... 7-16
    7.7.4 Moonlighting ... 7-16
    7.7.5 Working from home ... 7-17
    7.7.6 Corporate sector ... 7-17
    7.7.7 PhD plus skills ... 7-17
    7.7.8 Remote working to entrepreneurship ... 7-17

7.8 Plagiarism versus integrity; fake online reviews; publication charges; patent manipulation; citation multiplication; prolific authors, paper mills; fraudsters ... 7-18
    7.8.1 Plagiarism and frauds ... 7-18
    7.8.2 The peer review process ... 7-20
    7.8.3 Citation manipulation ... 7-20
    7.8.4 Words of caution to STEM PhD candidates ... 7-21
    7.8.5 Plagiarism, publication bias ... 7-21
    7.8.6 'Prolific' authors, article publication charges, predatory journals, paper mills, retractions ... 7-22
        7.8.6.1 Hyper-authorship ... 7-22
        7.8.6.2 Paper mills ... 7-23
        7.8.6.3 Predatory journals ... 7-23
        7.8.6.4 Retractions ... 7-24
        7.8.6.5 Scientific integrity ... 7-24
        7.8.6.6 Open science ... 7-25

7.9 Positivity in job search; networking benefits; workplace visibility; avoiding getting scooped; brain breaks; follow your passion; learning AI and ML skills ... 7-25
    7.9.1 Working to be employable ... 7-25
    7.9.2 Job search with a positive mindset ... 7-26
    7.9.3 Avoid getting scooped ... 7-27
    7.9.4 Keep relaxed, have brain breaks ... 7-28

7.10 Keeping the edge in employability ... 7-28
    7.10.1 STEM has an edge for getting jobs ... 7-28

		7.10.2 Continue to learn fresh skills	7-28
		7.10.3 Keep workplace visibility	7-29
		7.10.4 The key areas to focus on for STEM jobs	7-29
	7.11	A job in industry for STEM PhDs, or even a postdoc in industry	7-30
		7.11.1 Engineering remains a viable and important career path in industry	7-30
		7.11.2 EV profession as a case study for job growth	7-31
		7.11.2.1 H_2 fuel cells	7-31
		7.11.2.2 AI-supported self-driven cars	7-31
	7.12	Concluding remarks	7-32
		References	7-33

8 Growing to be a leader and staying on top — 8-1

8.1	Working towards a leadership role in a STEM job	8-1
	8.1.1 Getting noticed, gaining visibility	8-1
	8.1.2 Leadership through multidisciplinary talents for collaborations in a scientific career	8-2
	8.1.3 Developing skills for leadership	8-2
	8.1.4 Networking for leadership	8-3
	8.1.5 Collaborations	8-3
	8.1.6 When an offer comes, grab it	8-4
	8.1.7 Pay back with gratitude	8-4
	8.1.8 A position at the top brings administrative workload	8-5
8.2	Consolidating the leadership role to grow further, and to stay on top	8-5
	8.2.1 Stay positive while handling your group/colleagues	8-5
	8.2.2 Stand behind a junior who is being discriminated against	8-5
	8.2.3 Beware of gaslighting	8-6
	8.2.4 Learn to delegate	8-6
	8.2.5 Leadership during a crisis	8-6
	8.2.6 Remain loyal to your organization	8-7
	8.2.7 Share credit, as well as discredit	8-7
	8.2.8 Be one of them	8-7
	8.2.9 Inspire your team and people around you	8-8
	8.2.10 Be innovative and use modern and bold strategies as a leader	8-8
	8.2.11 Rising on the ladder	8-9
	References	8-9

Preface

During my long career since 1966, as a scientist, I have often observed how some PhD students, working in STEM subjects, do so well during their doctoral work, finish with flying colors and get employed quickly thereafter, whereas on the other hand, a good number of them struggle, with their doctoral work dragging. Is there a golden formula for success in doing a PhD in STEM successfully and making a career out of it? After all, every student starts PhD work with great enthusiasm and passion, knowing fully well that it may involve hard work of four to six years.

Apparently, to do a PhD well, what one needs is a strong thesis problem, a lab that has requisite facilities, a reputed PhD guide, a mentor to fall back on at every difficult stage during the PhD work, and, of course, hard work. Hard work implies collecting a lot of original data, and publishing at least 4–5 research papers in reputed journals, as the first co-author. But all this is more easily said than done. There are hurdles galore during the PhD work, and later, in the path of getting a job. What are those hurdles?

In the initial stage, PhD candidates are often made to do things that have nothing to do with the main theme of their PhD. This is done under different pretexts, but mostly to use the fresh candidate to do odd jobs. Even after the candidate has started his doctoral work, only a minority of supervisors find time to discuss the progress of that with the candidate, as he is only one of the many students being supervised by the same supervisor.

Efforts are hardly ever made to impart communication skills, both written and oral. Consequently, it becomes an uphill task for candidates to write the results obtained in the form of a manuscript to be communicated to a journal. Writing skills are needed at every step—while preparing a CV, writing the PhD thesis, or finally writing a job application, and a covering letter for it.

Public speaking skill is often lacking in PhD students. They definitely need it. It is a pity that a vast number of doctoral students in STEM subjects can't explain their own work in a simple de-jargonized manner, even at a stage when they have submitted their thesis and are ready for viva-voce!

During doctoral work, STEM PhDs generally dream only of academic placements after their PhD, such as a faculty position at a university, because they come across professors at their workplace who become their idols. Hardly any effort is made to make them job-ready for industry. It is a known fact that only about a fourth of STEM PhDs would find jobs in academia, leading to a tenured faculty position. The rest, after some postdoc work would need to look for a job in industry. However, only those candidates who have a good mentor learn how to secure a job in industry after a STEM PhD.

Employment opportunities for STEM PhDs have historically been much better than for graduates of other subjects. However, the Covid-19 pandemic and consequent lay-offs, and subsequent development of ChatGPT and a spurt in AI activities since 2022 have posed employment challenges for STEM candidates as well. Therefore, there is need for a book that can discuss useful ideas and approaches to tackle the hurdles faced by STEM candidates during their doctoral work, and provide directions for them to boost their employability. We do exactly that in different chapters of this book, by discussing how students working for a PhD in STEM subjects, anywhere in the world, can acquire essential skills and tune up to impress a job provider in academia/industry about their employability for the job applied for.

Key skills that will be covered include: writing, public speaking, self-management, risk management, collaboration, teamwork, conflict resolution, overcoming disruptions and developing skilful innovations at work to go for patents and entrepreneurship. The importance of 'networking' with experts in their fields will be brought out as a prime skill in getting a job, enabling them to prosper in it and be eminently successful later in a STEM career.

Communication skills to write a proposal for research funds and defend it before a funding committee are generally lacking amongst fresh PhDs from small, remote universities. In contrast, the success rate for proposals submitted by young researchers from well-known and established institutions is usually high, because they have opportunities to learn to write good funding proposals through discussions with colleagues in their big labs. To correct this non-inclusivity, an Indian funding committee for providing funds on organic electronics and biosensors, of which I was a member, invited fresh proposals/ideas for token funding. This was advertised well, particularly in remote and small universities. The applicants were asked to bring their draft proposals/ideas and make short oral presentations before a meeting conducted at Mumbai during May 22–23, 2008. Through mutual discussions amongst themselves, a majority of the candidates formulated and submitted proposals, 28 of which were approved for funding of INR 1.75 million each, with which a project-holder could buy a small instrument and employ a research student for a period of three years. This definitely sent a positive signal among fresh PhDs in STEM that it was indeed possible to write well-drafted funding proposals, get funds, and start work on their own research projects.

To boost their employability, STEM PhDs need to escape the trap of a 'single-subject cocoon' because the mastery of any single subject, such as physics, chemistry, biology, electrical engineering, computer science etc, will no longer be adequate to secure a job, in the near future. They will need expertise in two, if not three, non-overlapping subjects, and keep upskilling themselves to stay employable, and try entrepreneurship that would thrive often under a multidisciplinary eco-system.

To encourage a multidisciplinary interface, I organized a workshop, *Condensed Matter Interface with Chemistry and Biology*, during March 3–14, 2008 at Mumbai for over 40 young PhDs in condensed matter physics drawn from across India. The lectures at this workshop were delivered by top experts, including two from Japan, one from Italy, one from Finland, one from France, and five from Indian labs. Similarly, a course of four lectures, under the theme *Biology for the Chemists, Physicists and Engineers* was delivered by me at Bhabha Atomic Research Centre (BARC) in 2010.

Of late, the topics of my seminars have veered towards encouraging a multidisciplinary tilt among young PhDs. Some of my recent seminars were on themes:

(a) materiomics: from matter and materials to movement and life; (b) towards autonomous, self-propelled active matter; (c) convergence of art and science; (d) science drives innovations, and innovations create job opportunities; (e) bridging the gap between innovations and entrepreneurship; (f) what not to do to succeed in a research career; and (g) STEMM, AI and jobs.

People often ask me which areas in STEM have potential for new innovations, and chances to breed new technologies, and thereby create new jobs in the coming years. My answer is: (i) materiomics; (ii) living matter; (iii) homeostasis; (iv) wheel-free motion; (v) flying like birds; and (vi) microbiome. I further believe that potential developments leading to new STEM jobs will be in the healthcare sector. Making of a neuro-bionic interface, for instance, would restore brain functions without causing nervous system damage, so that neurosurgeons can cure a brain injury and move towards curing of diseases of the central nervous system among old people, such as Alzheimer's, Parkinson's, dementia, and epilepsy.

I am thankful to BARC, where I spent my whole career as a scientist, for having provided me opportunities galore to learn and practise science, and to hone my skills in multidisciplinary research.

I must thank the staff at IOP Publishing (UK) for extending all help during the planning of this book, and completing it. I got started in writing this book mainly due to encouragement from Ms Caroline Mitchell, the Commissioning Editor of ebooks. The ebooks coordinator Ms Isabelle Defillion extended support for a while, before Ms Betty Barber, ebooks Coordinator, took charge, and extended crucial help to me several times during the last year, as I worked on the different chapters, bringing this book into final shape.

Dr J V Yakhmi

Author biography

Jatinder V Yakhmi

Professor (Dr) Jatinder V Yakhmi (b. 1946), has spent a research career of 45 years at the Bhabha Atomic Research Centre, at Mumbai. Before his retirement in 2010, he was Associate Director of Physics Group, BARC, and Head of Technical Physics and Prototype Engineering Division. During 2012–16, he served as DAE Raja Ramanna Fellow at the the Homi Bhabha National Institute—a Deemed University of DAE at Mumbai. Simultaneously, during 2012–15, he worked as Chairman, Atomic Energy Education Society, Mumbai, which runs 31 Schools and Junior Colleges across India.

Receiving his PhD in condensed matter physics from Mumbai University in 1975, his focus of research has been on topics related to magnetism, superconductivity, and soft matter. He is credited with the establishment of a program on the use of molecular materials for fabrication of molecular magnets, sensors, bio-sensors, and organic electronic devices at BARC. Dr Yakhmi has a US and European patent on an artificial heart, and has contributed over 450 publications in international journals, including 65 review articles in journals/books. He has edited/written nine books, including *Thallium-Based High Temperature Superconductors* co-edited by A M Hermann and J V Yakhmi, published by Marcel Dekker, Inc., New York (USA), in 1994. Most recently, he has written a single-author book *Superconducting Materials and Their Applications—An Interdisciplinary Approach* published by Institute of Physics Publishing, Bristol (UK) in 2021. He has a Google Scholar *h*-Index of 54, and his name appears in the world's top 2% most-cited scientists, as per a list compiled by Stanford University.

Dr Yakhmi has delivered 150 invited seminars in reputed international labs, and about 50 at international conferences, and another 42 on topics related to popular science, higher education, and graduate jobs. He is a Fellow of the National Academy of Sciences (India), and an elected Member of the Asia Pacific Academy of Materials, and a Member of the Division of Condensed Matter Physics, Association of Asia Pacific Physical Societies. He is a winner of the triennial MRS-ICSC Superconductivity and Materials Science Award (Senior) by MRSI, Distinguished Alumni Award by Kurukshetra University, and the IIS Gold Medal by the University of Tokyo.

Chapter 1

Learning science: a passion for STEM subjects: why do a PhD?

In essence, science is a well-structured form of critical thinking, and a key driver of economic growth and social progress. Therefore, if science can be accelerated—such as by increasing the efficiency with which research dollars translate into discoveries and commercialized inventions—so can growth [1]. Academics have the rare privilege of pursuing projects that they find interesting and that they are deeply curious about.

Education in the subjects of science, technology, engineering, and mathematics (STEM) provides skills that make learners more employable. Science gives its students an in-depth understanding of the world around them, helping them to gain expertise in research and critical thinking. Technology prepares them to work in an environment full of high-tech innovations. Engineering imparts enhanced problem-solving skills and helps students to apply their knowledge in new projects. Finally, mathematics enables learners to analyze information, eliminate errors, and act decisively while designing solutions. In aggregate, the STEM education links all these subjects into a cohesive system, producing employable professionals who can transform society by innovating to provide sustainable solutions to problems at hand.

1.1 Scientific literacy

Studying science helps build scientific literacy, which means being able to reflect on science topics in your daily life, having skills to distinguish if something is fake and being able to discuss purposefully issues like climate change or even COVID vaccines.

Science literacy goes beyond just knowing scientific facts. When children learn science, they learn to question, reason, and deduce—all crucial elements of a critical mind. It encompasses a comprehensive understanding, appreciation, and application of how science works, its methodologies, and its role in our daily lives and society. In essence, science is a well-structured form of critical thinking.

Key components of science literacy are: (i) knowledge of scientific facts and concepts across disciplines; (ii) understanding the scientific method, which includes testing, experimentation, and analyses to derive conclusions; (iii) ability to think critically, which essentially implies becoming capable of choosing between credible sources and misinformation, and making qualified decisions based on evidence; and (iv) appreciating the role of science in society, ethics, and decision-making.

Studying science helps students understand what is happening around them, and helps them to look for evidence. It can help them, for instance, to take decisions whether, when, and where to go for solar panels.

1.2 Passion for science and STEM subjects

Most children have access to a camera, measuring cup, calculator, timer, ruler, magnifying glass, and GPS at home. Early experiences using such tools at home can help children feel capable and better prepared for learning science in school settings. When children have greater access to science tools and minor equipment at home, such as, measuring cups and kitchen scales—and time to tinker with them, it can strengthen their interest and confidence in science. The passion of children for science grows further when they have access to tools for building toys (like Lego and blocks), maps and compasses, telescopes and microscopes, and science kits.

The beauty of STEM education is that it inculcates creativity and divergent thinking alongside fundamental disciplines. It not only provides an understanding of concepts but also encourages application of the knowledge acquired by taking resort to two simple actions: explore and experience. Students of STEM get to learn from inquiry-based assignments, while retaining their focus on practice and innovation. Learners of STEM develop a special mindset for project-based learning and problem-solving. The core of STEM knowledge is flexibility and curiosity, which equips young learners to respond to real-world challenges.

1.3 Benefits of science education

By advocating pure science, we aspire to strive for scientific excellence. To achieve this target, our next generation of youngsters needs to be trained to acquire an attitude of being inquisitive, and always be ready to ask 'why?', even if it means challenging authority.

To do that, they need to possess the ability to do critical thinking, independently, and to acquire a speculative, imaginative, creative, and innovative mindset, with a determination to achieve their goals despite the challenges faced. It is worthwhile integrating these qualities into all our basic education systems—primary, secondary, and tertiary.

Obtaining a degree in the humanities is indeed an accomplishment, but a science degree is a lot harder to get than a degree in the humanities. You have to show up for class, whether for a lecture or for practical in the lab, do your homework and sometimes deprive yourself of pleasurable outings to get that project finished on time.

Students of humanities can come up with several different likely answers to a question asked in a class, but in science, there is usually only one answer, the correct answer, and you have to find it. Therefore, people with degrees in science and technology are valued a lot.

1.4 STEM education

As time goes by, it is likely that the teenagers of today will have multiple career options in their active lives and several changes of employers. It makes sense, therefore, for them to choose subjects that build skills which serve across several scenarios. Therefore, going for a STEM subject may turn out to be lucrative even if the career chosen is not related to science.

The core capability of higher education ought to be to build a capacity for critical thinking, and the concomitant requirement of evidence rather than assertion. That happens to be the essence of STEM—proof by experiment, mathematical derivation, and application of general scientific principles to a specific situation.

STEM education exposes students to effective interdisciplinary communication, and enables learners to make informed decisions within their subject areas. That enables them to find solutions for problems related to sustainable development.

At the foundational K-6 level, STEM education is limited to math and science curriculum that is required for all students. Hence, researchers into STEM education at the elementary school level focus on participation and the performance of students in science and math, in general.

However, STEM education gets defined more specifically as the curriculum becomes increasingly specialized at progressive levels of education. For example, in grades 8–12 multiple tracks become available to students through the required math and science curriculum, as do elective courses in the social sciences (e.g., psychology), computer science, and applied topics in engineering and technology.

1.5 STEM is the future

STEM research conducted by doctoral students at the university level led to the origin of several life-changing discoveries, viz. computers, antibiotics, transistors, lasers, FM radio, MRI, GPS, the Richter Scale, bar codes, Google, the fetal monitor, the nicotine patch, Buckyballs, nanotechnology, the insulin gene, bioengineering through the discovery of recombinant DNA, improved weather forecasting, cures for childhood leukemia, the pap smear, scientific agriculture, methods for surveying public opinion, the concept of congestion pricing, and human capital.

An integrated course of computer science, physics, math, chemistry, and biology, is useful to get into molecular and cellular biology, where the chances to get jobs are brighter. There are incidents when a student all set to do a major in English, and literature, decided instead to do computer science, mathematics, and astrophysics, with a degree in data science, knowing that as a field, STEM provided better opportunities to lead to potential jobs. In fact, since 2013 the number of students doing a STEM degree is soaring, as compared to those studying English and history [2]. In the US, the percentage of college degrees awarded in health sciences, medical sciences, natural sciences, and engineering has been rising.

President Biden has announced plans to stimulate US chip production. Apart from the economic benefits, the US is aware that much of the world's cutting-edge chips today are made by TSMC in Taiwan, the island to which China claims territorial rights. In the event of a conflict, therefore, semiconductor supply chains may be disrupted. TSMC's website shows that a majority of their top executives are PhDs, from top US universities like Yale, MIT, Stanford, UC, etc. The chips industry in the US, obviously needs a large number of students to become STEM graduates.

1.6 Why do a PhD?

Higher education is a reasonably sure way to secure professional opportunities and ensure the chance for upward mobility in a career. The PhD is a mechanism for developing high-level research skills, learning about rigours of its theme or the development of its theory. It sets you up with project management, problem-solving, and analytical skills that are meaningful within and beyond academia. During doctorate research one learns several transferable skills, project management, and even how to start a business, which are in essence, similar to doing a science project.

A PhD work is akin to a voyage made to understand how things work. It also provides a platform from where a candidate can be heard and can contribute to change. Doing a PhD provides an opportunity to dive deeply into a topic that the candidate is passionate about, and to contribute new knowledge to it. A PhD makes a candidate a critical thinker. The experience and skills acquired during the PhD work will remain meaningful for the entirety of their future career.

The business of a PhD Thesis is quite tough and involved [3]. After finishing your PhD work, you will learn, for example, how to analyze data. You will also understand how to examine your results to gain insights. In some ways, you are, in fact, better equipped than even MBA holders to make valuable contributions to the business world. You have learned resilience in the face of uncertainty and limited resources. As a PhD student, you probably spent more time 'doing' things, rather than studying or passively listening. As a PhD student, you conducted original scientific research, which may have included fieldwork, you taught younger students and you were involved with writing manuscripts. Science PhD students particularly benefit from near-constant immersion in emerging technologies and, especially, in data analysis and hypothesis testing. This becomes a huge advantage in the business world.

A PhD degree is the pinnacle of academic attainment for a young researcher. The holder of this degree is recognized as an independent researcher, an expert with in-depth knowledge of their chosen field of study, and a professional with a diverse set of transferable abilities. People who do a PHD quite often just love learning or are passionate about their topic. Life for a PhD degree holder can be much more nuanced, than without it.

Choosing a PhD problem is not a trivial task. Acquired relevant experience before a PhD is helpful for that, such as internships during or after a master's degree. Plus, you need to find a 'passion'.

PhD education goes beyond the classroom. One learns a lot through debates with peers, in dining halls and library foyers, and during writing for assignments. Open-access resources like the MOOCs cannot offer that.

PhD holders do earn more on average than their counterparts. Besides, doctoral education can empower and uplift, and open doors to learning of new skills.

1.7 Catch them young for STEM

It is common knowledge that students enrol in college to get a job after getting a degree. An 'employment imperative' is definitely pushing the new learning economy and changes in higher education.

STEM education is pivotal because STEM subjects decide the direction of humanity from generation to generation. Understanding how we affect the natural world is critical in creating both balance and sustainability. Science exploration and research provides the very knowledge of drastic changes that are occurring in the natural environment of the world's ecosystems.

If we want to raise the STEM aspirations of children we need to do it early, perhaps at primary school level or at middle school, before they are embroiled in the examination treadmill. By the time they reach their final years of school, pupils have generally already made up their minds.

When inspiration happens, education follows. Hence, there is a need to inspire and excite kids to explore STEM-related activities. Encourage young students to exercise their creativity and use STEM skills such as innovation, collaboration, and the engineering design process.

Former US President Barack Obama said: 'The quality of math and science teachers is the most important single factor influencing whether students will succeed or fail in science, technology, engineering and math. Passionate educators with issue expertise can make all the difference, enabling hands-on learning that truly engages students, including girls and under-represented minorities, and preparing them to tackle the grand challenges of the 21st century'.

Students of physics are generally better prepared for STEM careers, hence teaching of physics is essential in high schools.

Universities should help schools to ensure that children have an awareness and understanding of both the benefits and challenges of STEM education from a young age. By doing so, universities will empower young people to make truly well-informed decisions throughout their education. By doing it later we run the risk of fueling students' worries about the future and their place in it, rather than opening their eyes to the possibilities.

The real perks of a student attending university is that one gets to interact regularly with some very smart peers, which rarely happens in the outside world, and some such peers become lifetime friends. The 1996 Chemistry Nobel Laureate, Sir Harold Kroto stated once that only half of his chemistry knowledge was gained from text books and teachers at school and college, the other half was through animated discussions with his peers!

1.8 A PhD degree is a well-trodden path for employment

A PhD provides a very high level of expertise though in a narrow area of study—coupled with an opportunity to make an original contribution in that field. But none of that automatically translates into appropriate employment, either in or beyond the university, no matter how good you are or where you earned your PhD. Then how does it help in employability?

A PhD is a differentiator because it indicates (a) some minimal degree of conformism to work in a large group; (b) the ability to complete tasks; and (c) a minimal degree of being persistent over long time. A PhD degree confers knowledge which in turn confers the ability to understand complex ideas, and highlights the need for continued learning, so that a PhD holder can think critically and offer thoughtful responses to events or problems. A PhD degree does indicate your efforts to solve problems. A PhD holder is a person who can make long-term commitments and see them through to completion. This is a person who understands the necessity of meeting standards for performance. Thus, a PhD degree holder is quite well-equipped to do justice to a job in their area of expertise.

As discussed in section 1.6, one obvious response to why one does a PhD is related to better employment opportunities. A PhD degree is still a well-trodden path to relative financial success. In any case, higher education is currently in the midst of a transformation that goes beyond the vagaries of the job market.

Doing a PhD with an industrial partner can open up lots of career opportunities. PhD students sponsored by industrial partners end up becoming more interdisciplinary than someone who is just doing their project with an academic supervisor.

Years spent doing a PhD can be a time of discovery, meeting new focused friends, and learning new things. It helps young people interact with other students who may share much different life/educational experiences.

Young dedicated students deserve to do a PhD. They are a nation's future. Hence, let them not waste away their intellect, their talents, their inspiration, and their creativity. A PhD as an option will produce better outcomes for people, emotionally and financially.

1.9 PhD makes a difference

Why do a PhD? people often ask. More so, if it is not marketable in your profession. I had a PhD student who did a PhD in physics with me during 1993–96. By training he was a computer scientist, and after his PhD, he took up a job in the USA as a medical data handling executive in a hospital. In such a case, too, a PhD degree not only commanded respect among colleagues at his workplace, but the subject of his PhD—Physics, also stood him in good stead for his additional edge in problem-solving and critical thinking.

Doing a PhD not only teaches you how to learn, and be persistent, drawing upon your intellectual curiosity your whole life, a PhD degree represents the commitment to complete a long and sometimes difficult task. It will, among other things, make you a sharper thinker and a more critical consumer of information. It will make you less doubtful, a better communicator, and more reflective.

STEM covers a diverse array of occupations, from mathematicians to biomedical researchers. Some occupations like nuclear engineers do have a shortage of qualified staff with specific talents. STEM PhDs become a good alternate option, in such circumstances.

A PhD degree gives you polish and a set of shared experiences, an entrée into a shared culture. During lay-offs, employees with the highest qualifications in their subject area are affected last, obviously because the company wants to retain those with highest expertise and skills. If you are not laid off in your forties or fifties you will be thankful for your PhD degree. It will always be of value to your career capabilities. Let a PhD degree become more accessible and become a part of growing up as an adult.

The importance of higher education such as a STEM PhD goes well beyond the job market.

1.10 Passion for STEM—the lure of a PhD in STEM

Jobs made up of routine tasks that are easy to automate or offshore have been in decline. The number of jobs requiring greater cognitive skill have been growing. A skill that increasingly matters in finding and keeping a job is the ability (a PhD has it) to keep learning. When technology is changing in unpredictable ways, and jobs are hybridizing, humans need to be able to pick up new skills, using STEM knowledge. STEM PhDs, for instance, can do patent law and help inventors get their inventions patented by using their STEM skills (see chapter 5).

In many jobs, employers value communication skills, time management skills, collaboration skills, and critical thinking skills, and the skill to arrive at more diverse solutions to problems that come from exposure to diverse ways of thinking. These are all 'soft' skills that are the greatest benefit of a PhD training.

Every year, China produces more STEM graduates than the total of all USA students who graduate from college. Projections are that by 2025 China's yearly STEM PhD graduates will be nearly double those in the United States.

1.11 Why do a STEM PhD?

(i) To increase your salary potential. People who have doctorates are generally paid more money than those who don't.
(ii) To set a career change in motion. AI and automation are replacing many jobs pushing workers to reskill and upskill to remain relevant. We need to reinvent ourselves. A good strategy is do your research and try to predict what the in-demand roles will be in the future. Universities can actually help you here by teaching soft skills, in addition to knowledge, and prepare students for an uncertain job market rather than for specific jobs.

(iii) To follow your passion. Pursuing your passion is not a bad criterion for deciding to do a PhD. After all, people perform better and learn more when their higher education knowledge aligns with their values. If you can nurture your curiosity and interests rigorously, your expertise will be more likely to set you apart from other candidates, and increase the chances of ending up in a job you love. After all, even robots and AI are programmed to emulate this free-floating aspect of human curiosity in order to match human capacity for autonomous and self-directed learning.

It is less likely now than it was in the mid-twentieth century that a single paper or patent becomes 'highly disruptive', i.e. it changes the course of an entire scientific field. Science and innovation are drivers of both growth and productivity, and declining disruptiveness could be linked to the sluggish productivity and economic growth being seen in many parts of the world. This single reason deserves to push more candidates into doing PhDs in STEM.

Without doubt the work involved in a PhD does impart technical skills that are helpful for many jobs. Workers with a PhD degree are better writers, do better analysis, and create and articulate cogent plans. Many students who did PhD on a physics topic found jobs in the STEM industry where they apply their technical skills and also the soft skills learnt during their PhD work.

A lot of companies and organizations (including tech companies) need people with strong research, writing, and verbal communication skills. They even need people with artistic design skills. The best educated students are those who have a mix of liberal arts and pre-professional classes, such as those in STEM, engineering, or business who spend 30% plus of courses on the liberal arts. It is a myth that liberal arts have a monopoly on critical thinking and communication skills.

1.12 Employment after a STEM PhD

Focusing now on the 'job market', we can appreciate that there are two main tracks for employment of PhDs, viz. 'academe' and 'industry'. By 'academe' we imply tenure-track faculty positions, visiting professorships, and postdoctoral research posts; and by 'industry,' we mean full-time jobs in companies, non-profit organizations, government agencies, and everything in between. We will discuss these two tracks more in subsequent chapters of this book.

For many jobs in academia-land they simply demand a PhD degree, for instance for teaching philosophy, basic writing skills, research skills, etc to undergraduates. Universities hire mostly PhDs who largely end up doing that job for the rest of their lives. Many PhDs doing such a job do thrive, but hardly ever have to do anything that an MA/MSc could not do.

Doctoral students know that most of them will not find tenure-track jobs, yet 'assistant professor' remains the key career objective for the vast majority, and they are hyper-focused on the faculty path.

Employers outside academe, such as in select engineering and biomedical fields do base hiring decisions on an academic credential of a PhD degree.

1.13 Conclusion

In the era of artificial intelligence, modern technology needs quick solutions. Scientists, engineers, and business leaders often form a team to tackle that, and together they go through a daily routine of continuous learning. That is where acquired learning capabilities of STEM PhDs would help in creating a culture of learning to stay ready for new challenges. Besides, sharing of ideas and knowledge drops any sense of competition. For any organization or product team that is striving to solve grand challenges with technology, embracing these qualities as a team can help overcome setbacks.

STEM education will decide the future of jobs in machine learning, robotics, drones, advances in chemistry, medicine, and more. Encourage schoolchildren to do well in math and science. If they prove to have an aptitude for STEM, we need them to obtain electrical and chemical engineering and technical degrees. They will be in a great position to fill future jobs. Chip factories typically not only need technicians to run factory machines but also need scientists/engineers in fields like electrical and chemical engineering.

STEM jobs are growing almost twice as fast as other jobs. The enrolment level for science subjects needs to go up even as the demand for STEM skills is increasing. And that is so even in areas like climate change, materials science, health, food technology, drug manufacturing, and education.

Until recently, science was only useful for traditional science careers such as medicine or engineering. But that is no longer true. Many of those who do a STEM subject at school or university actually end up doing jobs outside of STEM.

The STEM-designated MBA, offered by some American universities, has been attracting a lot of students. A STEM MBA can open doors to global career opportunities, as many industries and companies require employees with both technical expertise and business skills.

References

[1] Clancy M *et al* 2023 To speed scientific progress, understand how science policy works *Nature* **620** 724–6
[2] Heller N 2023 The end of the English major *The New Yorker* https://newyorker.com/magazine/2023/03/06/the-end-of-the-english-major
[3] Ballenger B 2013 Let's end thesis tyranny *The Chronicle of Higher Education* https://chronicle.com/blogs/conversation/lets-end-thesis-tyranny

IOP Publishing

Employability for PhD Students in STEM

Jatinder Vir Yakhmi

Chapter 2

Sailing through PhD work successfully

Like all creative fields, doing the research work to earn a PhD degree in STEM needs dedication, determination, diligence, and tenacity, apart from a high level of acquired talents in one's own subject of research. Starting as a student of a Masters course in science, as one progresses up the ladder of success in a research career one comes across several challenges, some of which can be upsetting. Young researchers often need the support and understanding of a research guide, the group members in the lab, and family and friends, when they pass through 'low' periods, which is not infrequent. What are the 'pitfalls' to avoid in order to sustain a successful research career and help it blossom to its full potential? And how does one acquire academic resilience to face up to the challenges and sail through the work of acquiring a PhD smoothly?

After a short Introduction to doing a PhD in STEM (section 2.1), we shall first discuss the very basic and essential requirements (section 2.2) to be fulfilled, which if followed will lay a smooth pathway for the work related to doctoral thesis, followed by describing a few useful points (section 2.3). Together, the requirements and points discussed in these two sections are aimed at clarifying the important questions raised in the previous paragraph.

2.1 Introduction: doing a PhD in STEM for entry into exciting careers

A doctorate equips a young researcher to delve deep into the basics and acquire a keen eye for preparing a competent literature review, identifying useful research problems, identify research methods, collect reliable data and analyzing it to lay out the findings. A doctoral thesis underlines the significance of the research work proposed and completed. A PhD does imply an enhanced understanding of the research problem that was undertaken.

At university level, the percentage of staff with PhDs matters, apart from the fact that staff who have a PhD are the ones who are more likely to produce research, supervise other doctorates and contribute to the academic reputation of the department or the university.

Employers in industry, government, or academia find value in the training provided during PhD programs and find that with their depth of knowledge, intellect, and expertise STEM PhDs can contribute to greater public activities such as health-oriented programs, fighting disease, adult education, imparting better education to disadvantaged students, etc.

Increasingly, developing countries look at doctoral education in STEM as one more doorway to innovation, economic growth, and competitiveness. That calls for result-oriented research and mentoring environments during the PhD research work at university levels and large labs and collaboration with employers that hire PhDs, so that PhD holders can meet the current needs of the planet whether in the fields of climate, energy, hunger, or environment. This would obviously call for intensifying cross-sectoral collaborations with interdisciplinary research, entrepreneurship skills, internships in industry, and government organizations.

Currently, the US awards the largest number of STEM PhD degrees, followed by China, Germany, India, and the UK. In fact, China, India, and South Korea provide large inputs to human resources, i.e. students conducting STEM doctoral degrees in the US.

Which are the favored subjects for PhD degrees in STEM worldwide? These are engineering, biological and biomedical sciences, physical sciences, psychology and social sciences, computer and information sciences, geosciences, atmospheric sciences, ocean sciences, agricultural sciences, mathematics, and statistics.

Currently, PhD holders are expected to shape up as individuals who can 'address the planet's and our society's most urgent needs with imagination, humility and wisdom'. This calls for the creation of doctoral training programs that include preparation for multi- and interdisciplinary research, cross-sectoral collaborations, entrepreneurship skills, internships or secondments, and/or supervision in non-academic organizations.

Many developing countries link doctoral education to innovation, economic growth, and global competitiveness. For that to happen, however, they need to foster high-quality research and mentoring environments at universities, and collaboration with wider sectors of society as well as employers that hire PhDs. A good sign is that depending on the programs and priorities of a country, one sees an increase in funding for applied STEM research, which includes biotechnology, cybersecurity, and STEM learning and education.

Less than 2% of the population worldwide have a PhD. The myth about the unemployment of PhDs has been created by the media based on jobs granted to PhDs solely by academia. If one counts the PhDs employed in industry, government agencies, and business, then PhDs have the lowest unemployment rate in a nation's population. Then why do a number of PhD students feel lonely and depressed?

Doctoral research in STEM is a key to a strong base in science, that can spur the growth of a country and bring national prosperity, well-being, and international competitiveness. Lack of support for early-career researchers and inadequate supervision are causes of concern, which also relate to lack of training of senior researchers in mentorship.

In middle-income countries, only a few faculty members at universities have a PhD degree. Therefore, there is demand for doctoral graduates, with degrees from local and international institutions to add to the ranks of university academic staff.

Public media contribute to the myth of the unemployed PhD focusing solely on the academic sector and ignoring other labor market sectors such as government, industry, business, and non-profit. Considering the entire labor market and not solely the academic sector, PhDs have very low rates of unemployment. Nevertheless, we must not shy away from a discussion of the employment possibilities of doctorates, nor postpone the discussion by creating more postdoctoral (holding) positions without a career path available.

According to a book on doctoral education published by UCL Press (UK) in 2022 [1], the number of PhDs awarded in China, India and USA have been rising sharply. A majority of the PhD candidates are likely to get appointments outside academic research institutions, such as in industry, for which the PhD programs ought to be reformed to train the candidates to work across multiple disciplines.

2.2 Essential requirements for smooth sailing during doctoral work

2.2.1 Choosing a strong problem to do research towards a PhD thesis

Every aspiring PhD student wants to work on a strong and novel thesis problem, where there is ample scope to do original work. Experience tells us that important discoveries are made only when we study important problems, which are mostly tough to tackle. Easy problems lead to uninteresting results. Experts in their own fields of research often state that it is not enough that a problem should be 'interesting', because almost any problem becomes interesting if it is studied in sufficient depth. For that one needs state-of-the-art laboratory facilities, not just for experimental research, but also for theoretical research problems, where high-efficiency computation facilities can often bring home the bacon.

Picking a research problem of one's own choice is not possible when large research labs or universities advertise PhD programs based on pre-selected topics which fit into their scheme of ongoing research, for which they have funds to support PhD scholars. Additionally, working on topics pre-selected by these organization's topics is also likely to be smoother and yield results without wasting time because ongoing research programs would provide fair amount of lab facilities to work with, rather than waiting for equipment after joining as a PhD scholar.

2.2.2 Settle for a lab that has requisite facilities to conduct PhD work

Ideally speaking, a candidate should first explore the core strengths of a given department of a university or college where they commit to work for a PhD degree. This can be done directly, or by seeking the opinions of friends who are familiar with the existing research facilities, running projects or programs and research guides at that department. Then after some preliminary discussions with the proposed research guide, the candidate can settle to work on a specific topic for PhD under a broad subject in which a research program and the guide, both exist.

But candidates should also do self-assessment about their own capabilities to engage in research, because a PhD program is not just reading and writing what you like. Students who decide to do a PhD don't always understand what it means to do scientific research. Receiving good grades at university does not necessarily translate into being a good researcher. You also need to be lucky with your department and your dissertation committee. Succeeding in research not only draws upon a different spectrum of skillsets, but it also calls for a high degree of resilience and patience.

That doesn't mean panicking at the outset, because it is impossible to know what cards you will draw ahead of time, and you will have to learn how to play with them over the 4–5 years required to get a doctorate degree.

2.2.3 Try working with a reputed PhD guide/supervisor

Choosing an appropriate supervisor can sometimes be even more important than the choice of research topic. But what exactly is the role of a PhD supervisor?

Well, a PhD supervisor plays several roles during the PhD work of a candidate. A supervisor is not just a teacher who steers the PhD research work, but also plays the role of a mentor who can provide emotional support and facilitate the student's entry into a post-PhD career. A successful and efficient supervisor has a 'personality' endowed with several features, including a mentor, a guide, and a senior friend providing emotional support. In fact, a masterly supervisor would even caution the candidate to retreat when things are not going right during the PhD work.

However, not everyone seems to be able to supervise well, not because of lack of skills, but because it does not come to them naturally to provide supervision. Unfortunately, supervising a PhD candidate is not an inherent skill, and there is no formal training for how to guide and supervise PhD students anywhere in the world.

If you happen to make a bad choice of supervisor, then the chances are that you may have no-one to blame other than yourself.

Quite often you can do little in the matter since you are limited by your field of research and who will accept you. However, try making a choice among the available supervisors by talking to them, to their postdocs, PhD students, and even their former PhD students.

2.2.4 Avoiding some pitfalls while choosing a supervisor

It is advisable to have a supervisor who treats you as a 'colleague in training'. Someone who trusts your academic judgment, and doesn't make you do endless editing of your documents. A PhD candidate has to avoid the following pitfalls when choosing a supervisor:
- (i) You don't need a supervisor who already has a dozen postdocs and several PhD students, leaving hardly any time for a new student. Instead, you need someone who can spend time with you and shows interest in your success. Many reputed research guides are truly amazing scientists as well as mentors, but their popularity also puts overwhelming responsibilities and

obligations on them. Situations may arise when you can rarely meet your busy supervisor. Again, a high-profile supervisor, in all likelihood, will constantly travel to conferences due to their popularity, which leaves hardly any time for the PhD students, whose learning and research both suffer.

(ii) However, you shouldn't choose an inexperienced or easy-going supervisor who expects you to just repeat what a previous student in the group has done, with minor alterations, making the PhD problem very easy and comfortable, but hardly original. Hence, you need to avoid such supervisors.

(iii) PhD students should not become babysitters at the supervisor's home or be made to do other domestic jobs, even if you become friendly with your supervisor. PhD students are hard-working professionals, who are doing their best to survive, often with declining resources in the labs. They should not be loaded with onerous demands, totally unconnected with their research work.

(iv) You should not feel extra grateful for the bits and pieces of professional attention given by a busy supervisor. All that comes under the supervisory role anyway. Getting a PhD under a loose bond like that would reflect in low quality work or a lack of publications.

2.2.5 Collect a lot of original data for authenticity of results and reliable conclusions

To be able to reproduce one's data unequivocally, students must repeat their experiments 2–3 times or even more, in order to arrive at definitive conclusions. Statistically speaking, one must generate an ample number of datasets to be able to fit a useful profile to the data points to authenticate originality of conclusions being made.

A PhD project will unfold with time, in the matter of its precise details, viz. the role to be played by each partner. As the experience and mutual confidence grows, and depending on the nature and progress of the study, sudden useful results, if obtained can have a huge influence on the morale of the candidate.

Quite often negative attitudes of some lab-mates to score over other research workers by denying them resources may lead to unwanted delays in the collection of data, and completion of projects and a criminal waste of the precious time of young researcher, publication of whose research work gets delayed purely due to hoarding or not sharing things by such people. Hoarders never get any competitive edge, anyway, but one must stay cautious.

2.2.6 Publish as many journal publications as possible based on thesis work, of these, the majority should have the PhD candidate as first co-author

Try to check if prospective supervisors being suggested for you can be trusted that they do not hog the credit by publishing papers often with their name as first co-author, relegating the names of their students, even though they are the ones who worked to collect the data and analyze it.

During early days of the PhD work, a candidate is often told that they are not yet mature enough to earn a first co-authorship when a publication is being communicated based exclusively on the results obtained by the candidate during their own research. Excuses are given to explain why the supervisor or another PhD student, senior to the candidate, deserves higher credits owing to the expert help extended to the candidate. This is demeaning to the candidate.

Pressure to publish will continue even after a young researcher gets a regular job. They need to prove that they are bright and capable, for which the regular publication of papers, securing grants, and completing projects has to be done in a time-bound manner to retain and maintain their visibility in the research community. Therefore, it is appropriate to get into the habit of regular publications with due credits from the outset of their PhD work.

2.3 Other relevant points to be followed for unhindered PhD work

2.3.1 Acquiring of skills is important for doctoral candidates

We will discuss the learning of skills in some detail in chapter 3, and also in chapter 4, when we discuss the development of a multidisciplinary approach to gain an edge in employability during the eventual search for a job.

Doctoral students should acquire as many skills as possible, be they intellectual skills, academic and technical skills, or personal and professional management skills, which will help them to move successfully towards a PhD degree, and even beyond whether as a postdoc, or doing a job—in academe or in industry. Of paramount importance is to acquire the skills to teach and train students, and to participate in organizing scientific conferences and workshops, as well as learning to lead a group of young researchers; this comes through positive actions and treating everyone in the group with due consideration.

Early researchers working anywhere need to master skills to develop the appropriate strengths to successfully achieve their aspirations. Potential employers look for a broad set of skills, beyond academia, in a candidate who works for a PhD degree with them. This includes an innovative mindset, without compromising the freedom to do research related to the PhD topic.

A question often asked by job-providers from industry is whether the PhD candidate would be able to provide intellectual leadership, and has an appetite for risk-taking to drive projects towards the desired outcomes. There is no doubt that R&D spending drives innovation, but one just doesn't need to add more and more research jobs, at the cost of neglect of the economy and the labor markets.

Artificial intelligence (AI) has opened the doors for manipulation of large datasets to evaluate the complexities, or the inherent weaknesses of the system. It is expected that all professionals, including researchers belonging to any discipline will require these skills. The digital revolution has already led to the creation of jobs in industries which are digitally-intensive.

New PhD programs will need to emphasize the data and programming skills as an essential part of the training imparted to doctoral candidates. PhD candidates in all branches of STEM will need to build their capabilities and capacity to work with big datasets. Another important skill they need to learn is the ability to analyze the whole systems and how they interact.

2.3.2 STEM PhD does provide an edge

Doctoral training will prove to be a great asset and leverage for early-career researchers in STEM. They not only gain a multidisciplinary knowledge-base through academic interactions with other students and experts in the lab during their PhD research work, but they also gain useful core competencies such as resilience, stress tolerance, flexibility, active learning, etc. through social interactions in the workplace, which will improve their employability whether in academia or in industry, after getting their PhD degree.

There are tasks which cannot be automated (e.g. getting a hair-cut) because AI lacks the metacognitive edge that humans have intrinsically such as emotion in learning processes, regulating cognition during the process of acquiring knowledge, and above all the social motivation that drives learning.

It is becoming increasingly clear that AI when taken as complementary to human capabilities can boost productivity, but left alone, AI lacks the capability, yet, to automate certain key human capabilities, and some subtle human attributes like caring for others and understanding them, perception, manipulation, social intelligence, etc.

2.3.3 Learn to share facilities and consumables in the lab

Learn to share facilities and consumables in the lab. Some unscrupulous people hoard things for a 'rainy day', requiring other users to have to order materials unnecessarily, and lose precious time waiting for delivery while the materials are 'hoarded'. Practise a circular economy by sharing the existing materials in the lab and ordering collectively for common use by all. Sharing of resources invariably breeds positive collaborations between different research groups.

Big labs have huge stocks of chemicals, glassware, computer monitors, and other minor equipment piled up in storage bins or cupboards, all of which, if shared with the immediate users can drive more research with less funding required in the short term.

2.3.4 Be humble during field work

Fieldwork when required during a PhD time requires planning, resources, skills, hard work, and commitment. Learn from your mistakes, and accept the things you cannot change.

Having a modest view of yourself is essential to communicate on an equal level with people. As a scientist, your thoughts and concepts may be more valuable because you are well-educated and established, but you are the one asking questions —and the interviewees, whether they are fishers, farmers, or homeless people, often know more about many things than you do. Being aware of this is an expression of humility. I let the interviewees know that they were the local experts and I was the foreign learner. Interviews can take place in cafes, public libraries and picnic tables as 'mini-office' locations. In this way, one can recruit interview partners on the spot.

2.3.5 Need for a mentor

Your research program does provide you a formal mentorship structure, which includes your advisor and the principal investigator of the project under which you have been hired. Sometimes, these are the same individual, and may have several students to guide and in such cases will be a very busy advisor/mentor.

A mentor generally volunteers to advise a student, and clear up his doubts, even though they don't get paid for it. The student needs to describe their current skills and goals to a future mentor: whether they want to improve technical skills, build networking, or is looking for guidance for a prospective career in industry, academia, or government.

As a friendly guide, a mentor helps you to develop your career, and encourages you when you fail, or feel low. Having a mentor around is reassuring for a PhD student in the initial stages, when they start their research work in the lab where several professors, experts, masters or doctoral students move around, and the new entrant might feel uncomfortable. The feeling of being in a strange place can turn into a comfort zone with the right mentor at hand.

Thesis advisors, or supervisors, can themselves act as mentors, but not all of them have the time to spend on mentoring due to other responsibilities and obligations. In such situations, some candidates can be instructed in cohorts with more than one supervisor, so that a student is less isolated and better protected when their own supervisor becomes unavailable.

Graduate students sometimes build their own mentoring relationships either in the same lab or even outside it. Informal mentors could even be a past or current senior graduate from the same lab.

To build a career in academia, a mentor can help a PhD student in writing grant proposals; setting up collaborations, or solving the ticklish issues of co-authorships. On the other hand, a candidate seeking a job in industry may need guidance towards seeking internships, or learning additional technical skills. For candidates who are undecided about the course of their career in future, mentoring is a solution.

The mentor also benefits and learns from interactions with the mentee. Mentoring is a personal skill. Over the course of weeks and months they get to know the student and can counsel them on their main worries which may relate to problems with their thesis, the research facilities, financial aid, etc. A student performs much better and can accomplish a lot under the advice of a helpful mentor.

Doing a PhD takes 3–5 years, and during this long period there are always worries about tackling unfinished tasks. It is helpful then to depend on the advice of a mentor who is willing to help. Reaching out to a potential mentor means letting them know what you are interested in learning and asking if they would be willing to discuss it with you.

Quite often, people in a lab are afraid to speak up in meetings, or even to a senior leader, directly. Fear comes from an atmosphere of secrecy. It is better to grow in a transparent workplace culture where bullies can't operate. Mentoring is useful in such circumstances.

A mentor can be someone who is a postdoc working on a topic of your interest, or someone who has already received their degree from a lab or university you have joined or wish to join, or they could even be a researcher across campus who obtained a fellowship to collaborate with another university abroad. Finally, they could just be a scientist you met at a conference whose career path you admire. Peer mentoring too is a possibility, which is based on the reciprocity of a mentor and mentee co-learning and progressing together, and take decisive steps towards job-placement rates and social mobility.

2.3.6 Problems with mentorship

A mentor lacking the emotional intelligence to extend the much-needed support is worthless. Such a mentor fails to provide a positive perspective to the student, who may be in low spirits.

A favorite ruse adopted by unhelpful supervisors/mentors is to delay sending a research paper by the student for publication advising 'helpfully' that the manuscript does not yet look strong enough to be accepted, and why not collect some more data, etc. Then there is the 'super' mentor/advisor who wants to communicate the student's manuscript only to a top journal with very high impact factor, thereby attempting to boost their own CV, but decimating the chances of the student to get a publication in a journal of average impact factor. This can mean that the work is finally sent to a top journal, which rejects it, and thereafter the paper never sees the light of the day. Who is the loser, then? The student, whose career opportunities slump.

Poor mentorship can shatter a young researcher, delaying their career progress, as a consequence of which the student may quit their doctoral candidature, and in the worst case, may even leave STEM subjects altogether.

The student needs to remain alert, and discuss things, perhaps privately, with senior students or helpful faculty members on how to jettison a damaging mentorship.

In order to stay clear of the traps mentioned above, a PhD scholar should take stock of their graduate journey regularly, and make course corrections simply because the 3–5 years needed for a PhD is too long a time to stay compromised. It is true that for a successful career a PhD student has to work under the advice of a dissertation supervisor who is there to mentor you, give you advice you need when seeking a job, and even help you to network to improve your job prospects. But there can arise situations or variables which are totally beyond the control of the student, and any corrective action if not taken may ruin a budding career. Some students compromise by continuing in a poor situation, thinking that as they have already spent two or three precious years as a PhD aspirant, sacrificing a bit more might get them the degree. This luxury can be had by PhD researchers who already have a job, but not by students who will be seeking a job on the basis of a hopeless dissertation and an abysmally low number and quality of publications. I recall a case when the career of a PhD candidate had to be resurrected after two years of nil performance due to the non-cooperative nature of their supervisor. The candidate had to not only shift their PhD problem to another topic but also the lab where helpful mentors came forward to help and the candidate finally got a PhD, with a good number of publications after working for another three and a half years. But every candidate may not be this fortunate.

2.3.7 Attending conferences and networking boosts employability of a STEM PhD

There can be nothing more damaging to the morale of the PhD student than when their mentor, who is also the supervisor, forbids them to present their results at conferences. Not only does the student miss out on gaining presentation skills and professional confidence, but also loses opportunities to meet potential collaborators or industry leaders who control jobs.

A PhD scholar must attend academic conferences. That is where a young researcher can connect with new colleagues, make friends, impress the conference participants with their accomplishments, be introduced to experts, listen to and talk to them, and impress them to benefit their later careers. In fact, a fresh PhD has a good chance to land a PDF position by talking to the experts who can offer the same. The trick is to be on the lookout for the expert you want, seek their audience for a few minutes at the conference and explain your work very briefly, mentioning your key publications arising from the thesis work, and mention your ambition to work as a PDF in their group. A second method is to try to seek, during a coffee-break, the attention of the senior expert who asked you for a clarification during your presentation. That gives you ample opportunity to impress them favorably.

Social networking at conferences is an important opening for young researchers. If you can sit at the table occupied by a senior expert in your favorite research topic, that gives you the opportunity to explain your achievements in a nutshell and seek their advice on the career options that you have, and scope for future options.

2.3.8 Managing time is crucial for PhD students

Time management is important at graduate school, for productivity and efficiency. But for most students, time management is a challenge, because it is hardly taught anywhere at university level.

Time tracking can be of great help in achieving your goals as planned by injecting accountability to the amount of work done actually. Often students feel dejected that despite working for a full day they have not made much progress in working on the draft of a paper. But real-time time tracking can clarify how much of their time was spent usefully on the project, and what they spent on unproductive things like settling emotional issues, writing e-mails not connected with thesis work, looking at a distraction like the internet, or cribbing about the research paper that they are failing to finish writing despite efforts made.

There is a long-hours culture at many universities, including working through the night, with the number of hours put in at the lab climbing to more than 60 h every week. Long hours in the lab, with hardly any progress made, can often lead to burnout. Productivity demands focussed efforts, but overwork can also induce exhaustion, leading to impaired productivity. Obviously, satisfactory productivity requires time management, but that time management doesn't necessarily require cramming the maximum amount of work in a day; it should instead help you to prioritize your professional goals and structure your days and weeks without sacrificing personal well-being. That can help in cutting back a few hours in the lab leading to improved health, without a noticeable decline in proactivity. Reading, writing, collecting, and analyzing data, goes on non-stop as you continue your PhD work. But humans cannot work non-stop. Take a day *off*. Take a weekend off. Take a week off to recharge your batteries and return refreshed to take on the work where you left off.

Doctoral students should think about a PhD as a collection of a number of mini-projects. The best thing is to treat these mini-projects sequentially as per the overall plan of the PhD work-plan, and then allot them an overall time in such as over the next few months with detailed time lines of each subtask on a weekly or even daily basis depending on the size of the subtask.

Graduate students need to have long-term planning, working back from the expected time of obtaining a PhD degree, into which one needs also to fit the learning of special skills and also career exploration, maybe a postdoc position.

It is useful to work out a fresh plan of goals for the next quarter after the end of each term, after doing a self-assessment of time used effectively and goals achieved in the quarter gone by.

Finally, it helps to start the day by listing the tasks to be performed in a day. Or it may make more sense for you to do this only once a week since PhD research often involves experiments that run over several days. And you need to create special slots for some experiments which need to be done in a collaborating lab, to suit the time management of that lab.

2.3.9 Feeling lonely and depressed during PhD

PhD programs have, in general, the following challenges: a PhD program can take too long and be stereotyped in advising and even in choosing a PhD research problem. Often times, it exploits graduate students as cheap labor for teaching and for running errands. Worst of all, a PhD program is often in-built to secure a faculty position, and not for a job in industry, etc. Students need to be helped to apply their skills in a variety of workplaces, and not just in academe.

A PhD degree shouldn't be oriented just to produce a faculty job but should rather contribute expertise in the form of professionals serving every area of society.

The cost-of-living crisis is a fundamental threat for PhD scholars and early-career researchers. Financial distress could become an existential threat to PhD scholars. That turns out to be an existential threat to research itself. So, PhD students need to be paid appropriately.

Several labs in many universities have a long-hours culture, sometimes working even through the night. PhD students find it difficult to maintain a work–life balance. Some of them hardly get time off, their advisers sometimes even calling them on weekends. The intellectual stimulation and positive impact of peers can often get them through the PhD. Students often feel miserable, even though apparently, they appear to have made the right choice. Most PhD students have similar anxieties. They lack free time, they try to overcome similar failures, they rejoice like children when they succeed. They work a lot and most importantly, they have thoughts about giving up the doctoral work many times, but there is something that makes them *keep going*.

Doing a STEM PhD is a job filled with anxieties. After the initial excitement of joining as a PhD student is over, young PhD students gradually realize the hard job to be done. Some find that they are overworked and paid inadequate wages. And many of them feel the sudden separation from their families, which can continue for five years or so. Then, one has to worry not just about one's own experiments and thesis work, but also factor in how to get along with other people in your lab.

Some supervisors prefer quantity over quality in publications, which signals to the student that they might at best earn a poor-quality PhD, implying a struggle ahead to get employment as a PhD-holder. They additionally have to work hard to manage three things: teaching load, preparing for publications, and writing grant proposals for the group. Preparation of manuscripts for publication and grant proposals necessitates investment of large amounts of time on their part. Besides, they need to spend long hours learning new methods to conduct their research and do data analyses. There are increasing demands on the time of young PhD students to mentor younger masters degree colleagues, to attend to departmental duties, committee memberships, and to undertake work for professional societies. All this leaves little time for friendships, and they feel lonely. It dawns on young researchers how lonely research can be, when they find that there is a small audience for their presentation at a conference, which is uninspiring.

Being passionate about science is one thing, but working extreme hours for scientific research can be all-consuming. Risk of burnout is higher among young researchers [2], particularly among female academics from marginalized groups, because there is greater pressure to perform [3].

Even greater pressures affect young women belonging to minority racial groups, or those juggling motherhood with early-career research, those from the LGBTQ+ community, and scientists from countries where there is extreme gender discrimination or violent conflict.

Some doctoral students find it tough to learn the ropes of research and publishing, because just after joining they find themselves to be utterly unfamiliar with the unwritten expectations of the publishing methodology. Preparation of a manuscript, without much help from the supervisor, becomes a challenge. To complete their PhD work successfully they need to discuss and gain useful tips from senior students who may be at a stage closer to graduating, and a PhD degree. It would be a useful activity to become proactive further and form a supportive group of PhD students belonging to the same lab or the university to rely on one another and to learn and share strengths. Already successful friends may even share their favorite style of communicating research ideas and experiences. For instance, the art of storytelling, as a tool of communication, can turn out to be a winner, if rehearsed successfully among fellow-students.

Doctoral students undergo high levels of stress due to uncertainty in terms of graduate career outcomes. Before the pandemic, the general opinion was that about 20% of research students disengage from their PhD. Disengagement includes extended leave, suspending their work or dropping out altogether. Covid-19 has only added further to their disenchantment.

Working for a doctorate is a transformational process evolving constantly over a period of four to five years, generally, which involves intellectual orientation. A PhD degree brings high expectations. But it also brings highs and lows of emotional stakes, which can be daunting. Students with low resilience levels often suffer from anxiety, depression, and trauma. A useful suggestion is to divide the overall research work into small, clear tasks, which are doable and manageable, like climbing a ladder step-by-step. To do that the candidate has to first grasp the overall task in its completeness in consultation with the PhD supervisor. Once enrolled as a PhD candidate, one has to periodically, perhaps once a week do a self-analysis on how the PhD work is shaping up, even if bit by bit. The candidate has to ensure that the orientation of the work towards the final goal stays intact. This may need some improvising, at times.

All PhD students need to do two things extensively: *read* and *write*. It is advisable to get into that habit as early as possible. In order to write well, you have to read a lot, and also learn to distil useful information.

Deadlines, failed experiments, pressure, competition, fatigue, a feeling of dejection, obligations, are all there to remind us of the *value* of what we do, why you are doing a PhD, and what is your ultimate goal. Cultivate a supportive network of colleagues and friends. They may not only help you with your research, but they may also provide *emotional* support when you need it.

Serving on committees and other extramural distractions should never be used as an excuse for not doing research, for that is the scientist's first business.

There is a perception that PhD students are a privileged lot. Too often, nobody bothers to really ask them about their feelings and frustrations. In graduate school, you too would think you are self-sufficient, working intelligently and methodologically on a problem, but there are continuing adversarial interactions: the endless conflicts of peer review, disagreements with faculty, passive-aggressive questions at conferences. It is very important for PhD students to have an outlet for this.

Many PhD students are in distress. About half of them are studying away from their home country. And if that is not enough, tackling a new field of research can be a source of anxiety. Having to admit that you're struggling carries a stigma. Some of them even decide to quit their graduate studies.

The top two sources of emotional strain are the uncertainty about job prospects and difficulty in maintaining a work–life balance. Lack of funding, and inability to pay off debts is also a major concern. Some PhD students also have families to support—a huge challenge. Some have to spend on day-care centers to look after their kids. Being a good parent and an efficient PhD scholar doesn't leave any time for leisure, because most of the time it is either PhD work or parenting.

Some PhD scholars undergo a state of mind that transcends qualifications or accomplishments—they feel that the more they learn about something, the more they realize that there's a lot more to understand. No wonder their satisfaction levels fall as they dig deeper into their PhD program. Not being able to meet their original expectations can be a major source of dissatisfaction and disappointment.

2.3.10 Be clear and selective while making commitments at the workplace

If a research task doesn't contribute as input to your future publication, then you might take it off the list. PhD students should learn to say NO, if the task being asked of them is time-consuming and of absolutely no consequence to their PhD project. Otherwise, they can commit to too many useless things which eat into their valuable time, which could be more gainfully utilized towards achieving the main goals of their PhD work. Let it be known that you are not easy meat and you are not available for useless, petty errands/jobs to be thrust upon you.

On the other hand, you also must not give the impression that you would say NO to any task allotted, lest that makes you miss a chance to connect with someone useful for your thesis work, or gain some useful experience by doing a short task, even if not connected with your PhD problem. Therefore, be selective and considerate before saying NO. This way, you will not be branded as non-cooperative, and at the same time, get a chance to expand your horizons or gain useful connections, by saying YES once in a while to doing a task.

In a recent job interview at a biotech company, a biology graduate was asked if they were able to operate drones. It wasn't part of their university biology curriculum, but having flown drones as a hobby would have landed the job. It's this sort of lateral skill that can provide the extra advantage that makes you stand out among your peers. You rarely know what will be useful. The future belongs to those who can combine many skills, any skills.

In some university systems, the PhD can be misaligned with the requirements of a scientific career. Not much thought is given to the research project related to the PhD, resulting in many students graduating without any publications to their names. This creates stress and difficulty anticipating the hardship in navigating the next phase of a career, while applying for jobs or fellowships.

Students are often advised to register for a PhD program as soon as they become eligible. This may not be a helpful advice, and students should challenge it, and/or initiate a discussion with their mentors on this issue. The student needs to weigh up, keeping in mind that one needs to work hard to get a PhD, whether this means spending hours standing by your research equipment or days and nights in front of the screen of your computer. 100% dedication and focus on the same topic have to be maintained for 3–4 years, which is the typical duration of a doctoral program.

2.4 A PhD degree should lead to careers and jobs

A general concept is that a PhD will dramatically improve one's job prospects. Perhaps due to living in a research-based academic atmosphere as a graduate student a majority of the young doctoral students dream of a career in academia, the rest have a fancy for jobs in industry, knowing that technology is changing, and consequently also the job market. The employability aspect is important and needs to evolve with the active interest of all the stakeholders including the early-career researchers who recently achieved a PhD, academia, and industry leaders [4].

2.4.1 Careers in education research

A postdoctoral position in education research should be encouraged, which helps the candidate use their research expertise in teaching, and may lead to a tenure-track faculty position with expertise required in education research. Such positions have started coming up in science departments at leading research universities across the world, although they are not yet very common. Faculty members in STEM education research should have the option to keep a strong research program that focuses on education, while continuing with their teaching programs.

As the number of faculty positions focussing on education research grows, early-career researchers will be able to enter education-based STEM research directly. This will have far-reaching benefits, because faculty members who focus on education will not only support their own students, but also help their colleagues to adopt the latest, science-based teaching methods, which in turn will improve the reputations of universities with applicants.

The US National Science Foundation funds extensive research into science pedagogy, much of it through its Directorate for STEM Education, which provides US$5 million to postdoctoral fellowships in STEM education [5].

PhD students joining conventional subject-based programs are often book smart, but lack problem-solving skills, which are needed for instance in a physics research labs to interpret results or master the use of scientific instruments. Undergraduate education of 'technical expertise' in physics, engineering, and medicine, is therefore of great relevance. Postdocs in education research, can enrich the interface between scientists and school classrooms, and can bring together actual STEM researchers and classroom teaching. Manuel João Costa, deputy rector for student affairs and innovations in teaching and learning at the University of Minho in Braga, Portugal, advises following the social-media accounts of teaching and learning centres, which often post content of interest.

Education-focused STEM academics can benefit both students and faculty members in science departments, because they understand their science discipline better than scholars who are trained in education only.

References

[1] Nerad M, Bogle D, Kohl U, O'Carroll C, Peters C and Beate Scholz B (ed) 2022 *Towards a Global Core Value System in Doctoral Education* (London: UCL Press) https://uclpress.co.uk/products/176625, or go.nature.com/3zihyuk

[2] Boone A, Vander Elst T, Vandenbroeck S and Godderis L 2022 Burnout profiles among young researchers: a latent profile analysis *Front. Psychol.* **13** 839728

[3] Kucirkova N I 2023 Academia's culture of overwork almost broke me, so I'm working to undo it *Nature* **614** 9

[4] Clancy M *et al* 2023 To speed scientific progress, understand how science policy works *Nature* **620** 724–6
[5] Dance A 2023 Classroom assistance: the scientists turning the tools of their trade to education *Nature* **613** 203–05

Chapter 3

Using PhD final year and just after to master useful skills—writing and public speaking

3.1 Role of effective communication in science is crucial for a STEM PhD

Communication of any kind means sharing knowledge. Communication in science, whether oral or written must have accuracy, brevity, and clarity, to be effective. While communicating in science it is not enough to think only of the topic and the message to be delivered; one must also consider the frames of reference of the audience/readers and the questions they might have on the given topic. In fact, a presentation should induce and tempt the audience/readers to ask questions.

Effective communication bridges the gap between the knowledge and interest of the audience and the content of the document or presentation. PhD students will perform a lot better if they can learn communication skills that will serve them well in all sorts of careers, both in academe or in industry.

When your audience is less specialized, the gap is wider and bridging it is harder. While writing or speaking specifically for non-specialists, remember to use analogies, provide visual representations (with an idea of scale), and so on.

When writing or speaking for a strongly heterogeneous audience, include first what everyone is primarily interested in and later what only some of the audience needs or wants to learn. In all cases, do not overestimate the audience's knowledge of your topic or field, but respect their intelligence, all the same.

One big difference between academic and popular science communication is vocabulary. Technical jargon can alienate an audience of non-experts, who have little formal knowledge of the problems under discussion. It is a good exercise for STEM PhD students to learn to describe their research separately in just about 200 words, separately for (i) a general audience, which has little science background; and (ii) the academic community. This will pay dividends during multidisciplinary collaborations, such as between engineers and physicists, essential for solving complex problems, or for physicists while contributing to a remotely connected field such as, biomedicine. Innovative partnerships between subject experts from different fields will emerge only if research is communicated with clarity, in simple terms, to a broad range of potential collaborators.

The essence of a PhD research work spanning three years, or more, of independent study mentored by a supervisor is recorded in a doctoral thesis, often a book-like piece. The work explained in it is assessed in an oral examination by a few senior academic researchers, during viva-voce.

A point to be noted by STEM PhDs, when they search for jobs, is that the majority of them will find employment outside academia, hence, they should work on their presentations accordingly. Professional doctorates, often done in engineering subjects, are often, jointly supervised by an employer and an academic, and are aimed at solving industry-based problems. Therefore, it is preferred that there is a close partnership between education researchers, PhD supervisors, and organizers of doctoral-training programs in universities for effective liaising with industry for securing jobs for STEM PhDs.

A uniquely innovative way of doing PhD is by publications, though it is not so common. For this, instead of writing a thesis on one or more research questions, the criterion for an award is a minimum number of research papers published or accepted for publication.

Although, we shall elaborate in section 3b on the usefulness of attending conferences for PhD students, it is tempting to suggest participation of doctoral candidates in Lindau Nobel Laureate Annual Meetings, which are held in Lindau, Germany, mostly in June. Each year, about 30 Nobel laureates come to Lindau to meet between 400 and 500 undergraduates, PhD students, and postdoc researchers from all over the world. Journalists from several countries are invited also. The goal of the meetings is to foster exchange among scientists of different generations, cultures, and disciplines.

Considering the importance of the topics to be discussed in chapter 3, we have divided it in two parts, 3a and 3b, as follows:

Chapter 3a. Writing of manuscripts, CV, resume, thesis, proposals for funding and applications for jobs

Chapter 3b. Public speaking: presentations at Seminars, conferences, before funding committees, job interview committees, and viva-voce

3a.1 Learning to write on research topics: avoiding plagiarism

Written science communication takes many forms, such as manuscripts for publication in journals, reports, conference abstracts, review papers, theses, research proposals, popular science articles, and poster presentations. These various forms have a lot in common, but they also differ in purpose and their readership.

If you want to be a good writer, you must read a lot and write a lot. There's no shortcut to that because you will learn most by observing the work of other writers, who are experts in it. Before communicating a manuscript for publication, you as the author must ensure that it is clearly written; the data reproducible; and the data and message is new to be considered as contributing significantly to scientific knowledge. In addition, the message or the conclusion being made must arise from the data being presented.

Scientific papers, reports, and theses usually follow a standard format, incorporating different sections: Abstract, Introduction, Materials and Methods, Results, Discussion (or Results and Discussion), Conclusions, and References. To be worthy of consideration by a high-reputation journal, a manuscript ought to be comprehensive and focussed, and should highlight the newness of the observations, apart from bringing out the essence of the achievements made. It should also mention the possible impact of the results obtained on their applications in society, or industry.

A thesis has an additional section—Literature Review. Similarly, a Review Article or a Book Chapter have their own formats.

Technical writing is not just simple writing but rather a unique and very distinguishable skill. It can be learned and mastered, but it requires time, hard work, and practice. However, once developed, the techniques of technical writing are useful not just for preparing technical presentations, but also for drafting applications for research grants and scholarship applications.

It is crucial for scientists, engineers, and citizens to be able to critique data, to identify whether or not conclusions are supported by evidence, and to distinguish a significant effect from random noise and variability.

As a technical writer, you must always strive for clarity, conciseness, and coherence. If you keep that in mind, you will succeed in presenting the complexity of your research in a way that makes it easy for the reader to understand the concepts, get the key information, and also give you the credit you deserve. Organize each section of your text in paragraphs. Each paragraph may have 10–15 lines. Keep the sentence structure simple to retain clarity and readability.

3a.1.1 Academic integrity

Academic and scientific integrity relates to knowledge, skills, and values. While quoting results from literature, it is essential to provide the correct reference and give credit to authors. It is essential to acknowledge the original source properly, and disseminate scientific findings with transparency and accountability.

After joining a PhD program, a candidate will have to learn to maintain scientific integrity by avoiding the pitfalls of misconduct and malpractice. And learn also how to protect the value produced by science (intellectual property), and finally, to communicate science keeping ethics in mind.

Before we discuss how best to write the manuscript of a research paper, it is pertinent to discuss how to avoid plagiarism. Not doing so can lead a young researcher into lifelong difficulties and even a loss of job.

3a.1.2 Avoiding plagiarism

Borrowing without proper acknowledgment is a form of dishonesty known as plagiarism. The advent of the internet as a resource for research and information has opened a path in which students, and even their teachers, can easily, perhaps unintentionally copy information belonging to other authors. The 'cut and paste' facility has opened up a wave of unintended plagiarism. Although software that can detect plagiarism is now available, it should not be used to circumvent plagiarism. Knowledge and ethics should be the sole defense. Students should know that they violate academic-integrity standards by indulging in such activities. We have the freedom to use the work of others in our own work, as long as the appropriate attributions are given. Not doing so is the equivalent of stealing. Rules for attribution and reasons for their existence are rooted in honesty, integrity, and respect for others [1]. We need to demonstrate our understanding of copyright in everything we present, as an act of integrity.

PhD holders should identify plagiarism and understand self-plagiarism. Self-plagiarism is defined as a type of plagiarism in which the writer republishes a work in its entirety or reuses portions of a previously written text while authoring a new work. Writers often maintain that because they are the authors, they can use the work again as they wish. But self-plagiarism can infringe upon a publisher's copyright. Authors and researchers must take preventative measures in their writing practices and editing processes, to stay clear of potential self-plagiarism before submitting their work for publication.

Self-plagiarism also includes the efforts by authors to publish new papers after modifying their published results, slightly. This subtle form of plagiarism, too, is unethical, and so is 'salami slicing', where many similar papers are 'sliced off' from the same dataset, primarily as an attempt to bolster the number of publications.

Graduate students and junior scientists often add the name of their advisors or project principal investigators (PIs) as co-authors of their articles. Termed ghost or gift authorship, this practice is unethical, and formally prohibited when the senior author does not contribute to the publication. Adding fake co-authors from prestigious foreign institutions, confers some distinctiveness to their work, undeservedly, which is unethical.

The 'publish or perish' incentive drives many researchers to increase the number of their papers at the cost of quality. Research quantity and existing citation-based quality metrics, such as the impact factor and h-index, are easy to measure. Such measures are, therefore, often used for key decisions such as funding, tenure, and promotion when large numbers of competing researchers need to be quickly compared. However, no metric should be taken as a substitute for the actual reading of a paper in determining its quality.

3a.1.3 Copyrights

All creative works are protected by copyright. A copyright, is the legal right granted to an author, a composer, a playwright, a publisher, or a distributor to exclusive publication, production, sale, or distribution of a literary, musical, dramatic, or artistic work. Copyright laws are based on the belief that anyone who creates an original, tangible work deserves to be compensated for that work. Just quoting or crediting the author of a copied work does not satisfy copyright requirements.

Copyright guidelines do not allow users to make multiple copies of different works as a substitute for the purchase of books or periodicals; copy the same works for more than one semester, class, or course; use copyrighted work for commercial purposes; or, use copyrighted work without attributing the author.

3a.2 Writing of manuscripts to publish in a journal

Start writing a paper only when you have robust results from different experiments that support your idea, and a good overview of how your research advances scientific knowledge. Separate the writing into parts. Let the results and the methods be written first, since that is where you write what was done and how, and what the outcomes were. Next, tackle the Introduction and refine the Results section with inputs from your supervisor and collaborators on how to develop the story, references to be included, and the takeaway message. In the last part write the abstract (also a graphical abstract if required). Finally, work on the conclusion. To elaborate:

The **Abstract** should be clear and concise, after reading which a reader can decide whether to read your paper in detail. The **Introduction** should describe the motivation for the work, and the key findings in some detail to support the same. The **Discussion** section should focus on the main scientific and societal takeaways from the research work being reported, keeping in mind the specific interests of the journal readership.

The target journals, too, sometimes dictate the format of writing and preparation of the manuscript, particularly the figures or dataset, word counts, and reference formats.

If you are working on a multidisciplinary project with co-authors who were involved in the project planning, then have their inputs earlier in the process to ensure that the storyline is clear and the interpretation is sensible to multiple disciplines. Consult your co-authors and collaborators also for the manuscript structure and about suitable journals.

Do not limit the circle of people who provide comments on your manuscript to direct collaborators. Feedback from a colleague in a different discipline can help you identify issues you might otherwise overlook. A good idea for excellent research needs to be supplemented with good, and clear writing. Before submission to a journal, it is prudent to let a colleague or two pre-review it. Let it be seen preferably by two colleagues: one who is familiar with your research work, and the other who knows hardly anything about it. The former can provide technical advice, while the latter can determine whether your ideas are being communicated clearly.

Send your manuscript to the right journal. Many rejections result from a mismatch of a manuscript and the journal's scope or mission. The covering letter should preferably suggest the names of reviewers for your manuscript, especially considering when the editor isn't well-versed in the subject of your manuscript.

In the long journey of a PhD studentship, the students need engaged supervisors, who are available to interact and advise them and inform them of potential pitfalls which would help them to make informed decisions about their PhD journey, including writing and communicating manuscripts for publication.

Publishing in a journal is an integral part of being a researcher. A published paper can be read by researchers around the world, and allows other researchers to see how the research was done and maybe question the results. A publication is a permanent record of what has been discovered, when, and by which scientists. Finally, it not only shows the quality of the researcher's work, but also helps researchers to promote their work and gain recognition from funders and other institutions, including funding authorities.

The key to successfully publishing an article is to have a vision—a reason, a purpose for writing. Without vision, the research paper may not see publication in a journal. It is often possible to target your research work for a specific journal, by deciding which components of the study to put into that manuscript, what types of analyses you want to present, and which ideas you want to emphasize in your paper. Writing your paper specifically for that audience of editors is a key strategy in getting your paper accepted. If the results being reported suit any particular scientific specialty, then a journal pertaining to that might be appropriate. Further, check if the present work fits into a theme that the journal has been pursuing.

A pre-submission inquiry can be made to gauge a journal's interest in your work to be reported. For that send an extended abstract listing the purpose of the project, methods, main findings, and conclusions. This abstract should be written for non-specialists, and it should include citations of relevant journal literature. A covering letter should briefly describe the theme of your research project, what you did, and why your results are significant enough to be of interest to the readership of that journal.

After the choice of the journal to which your manuscript is to be sent has been made, read carefully the journal's editorial guidelines, and follow them carefully. The abstract of your manuscript should highlight the main results.

If you are the first author you should write the first draft of the paper and you should prepare the figures, tables, and legends. Then let other authors work with you to get the paper into shape. Send that draft to colleagues in your field and department for review and comments. Lastly, let someone in your lab proof-read it.

Don't panic at the non-acceptance of your paper, because it doesn't always mean a rejection. Almost no manuscript is accepted without any revisions. Only 5%–10% manuscripts are accepted the first time they are submitted. Most other acceptances are subject to revision. Mostly the responses from a journal are 'make some minor changes', or 'revise and resubmit', or 'reject and resubmit'!

A manuscript is rejected only for one or more of three reasons: (i) it is not within the area of coverage of the journal; (ii) it has serious flaws in methodology, analysis, interpretation; or even in the main question being evaluated itself; (iii) it has very little novel contribution to the field.

When does a manuscript belong to the 'poor' category? When it is seen to be deficient in review of the literature; has irrelevant citations; its introduction is not meaningful; its research proposal is faulty; methodology is insufficient; samples studied are not described adequately; analysis of data is unsatisfactory; discussion is written poorly; writing style is not of high enough standard; and/or it is undesirably long.

The decision to publish a manuscript is taken by a research journal normally after a peer review process, in which the quality of a manuscript is evaluated by one's peers in the scientific community. Authors can expect their manuscripts to be reviewed fairly. We discuss more of the peer review process in section 3a.2.1.

Before we discuss the peer review process, let us once again go through the purpose and focus of different components of a manuscript.

Title of a research paper. A title is composed usually of not more than 20 words, in which the authors must convey the main idea and the contribution made. It states what the paper is about, and speaks about the science in it.

The abstract. This should largely convey the essence of the study and the main achievement made under the study. It is a window to the content of the manuscript. Most readers do go through the abstract of the published manuscript to decide if they wish to read the whole article. An abstract written well conveys the research proposal and the main findings, and can induce the readers to read the whole article, if they wish.

Introduction. The introduction provides a context for the study made, presenting a rationale for it, in the context of current knowledge and prior theoretical and experimental work on the topic. It underlines gaps in knowledge of the research problem being handled and states the research question, its purpose, and how the study has been planned.

The method. This section describes the details of how the study was made including the apparatus used, procedures, and the extent of data collection, in a clear manner, such that any other researcher could duplicate the study, if desired. In particular, the characterization of the samples used has to be described in a transparent manner.

Results and discussion. The results section incorporates a summary of the collected data, its analyses, and the plan of analyses. All results should be presented, even if they include unexpected findings. In the discussion section, the authors present how the results were evaluated and the main findings made. Limitations, if any, should be discussed and the importance of the findings presented. The authors may offer recommendations for further study.

Tables and figures. These should add value to the presentation of data and analyses. Relationships among data should be highlighted. In order to avoid duplicate reporting of data, the authors should choose the most comprehensible methodology of presenting the information, which could be text or graphics and/or tables.

All through the manuscript, be correct, exact, unequivocal, and grammatically correct in your writings. Be concise and avoid repetition. Make a small section of acknowledgments recognizing the help received from people, not mentioned in the main text.

3a.2.1 Peer review

When a paper is submitted, the editorial office of the journal looks at it briefly to check the format and length, the clarity of the discussion, research methods, and overall fit with the policies of the journal. If it passes these criteria, then it is sent for full review to subject experts, i.e. for peer review.

Peer review is the system used to assess the quality of scientific research before it is published. For this, independent researchers in the same field scrutinize research papers for validity, significance, and originality to help the editors decide whether or not a research paper should be published in their journal.

Peer review not only provides a system to select which research should be brought to the attention of other researchers, but it also gives authors feedback to improve the quality of their research paper before publication.

In the process of doing peer review, the reviewers develop their own skills in conducting research, writing, and data presentation skills, and their ability to evaluate their own work objectively.

In traditional, single-blind peer review, scientists do not know who critiques their papers. In double-blind review, neither the study authors nor the reviewers know each other's identities. In double-blind review, real names are attached to the published paper by the editor only just before publishing it.

Most publishers of academic journals make a profit from subscriptions and sales, but the peer reviewers extend their services on a voluntary basis, i.e. for free.

Peer review is a system in which journals have no way to coerce reviewers to return their critiques faster. To spread the network of peer reviewers wider, there are suggestions that postdocs or doctoral students in their final year of PhD work, should be enlisted as reviewers, keeping in view that some of them may indeed have the required expertise in the subject area of the manuscript to be reviewed, and also that it may in turn sharpen the much-needed analytical prowess of the doctoral students.

Finding subject reviewers is a careful procedure because it is voluntary and anonymous. An Editorial Board uses its database of previously published authors and reviewers to decide on which researchers should be asked to do peer-reviews for a given manuscript. Some journals welcome the authors to suggest possible reviewers.

The peer review process is also utilized by universities for a thesis received for award of a PhD degree. The same is true for a review of funding proposals.

Reviews are kept anonymous to protect reviewers and allow them to give honest feedback. Except that the reviewer for a PhD thesis can be easily found out, but then the rate of rejection of a thesis is perhaps almost nil, whereas for a funding proposal it is 10%–80%, and for a research publication it is about one out of four proposals, at most. This is also to protect the data to be replicated quickly by competitors

Anonymity allows constructive criticism, and also protects a reviewer from retribution for a negative recommendation, also from sexism or racism. The *British Medical Journal* publishes names of reviewers if the paper is accepted, but not if it's rejected.

3a.2.2 Committing fraud, paper mills, and the after-effects of retractions

There are many incentives to publish in the research sector, pushing some scientists to cut corners and lower their standards. Publication is a principal criterion for career advancement in the research sector. STEM PhDs shouldn't salami-slice their results into three or four separate publications, rather than one meaningful publication. The overall quality of research is strengthened if researchers follow research integrity practices.

Some scientists buy papers from third-party firms to help their careers, ghostwrite papers on fictional research, or bypass peer review systems for payment. The overall size of the **paper-mill** problem probably runs to thousands or tens of thousands of papers. Funding agencies have to be careful about where papers are published.

Reputed publishers have teamed up with research groups to develop software that can detect duplicated images across published papers. Software is improving but isn't yet capable of looking through papers on a massive scale.

The number of retractions and their continued growth in numbers tells its own story. More often than not, the story involves misconduct or fraud. Therefore, no reputable publisher goes to press with a manuscript today without first running it through a program to identify plagiarism.

Peer review also has a drawback. It offers the greatest possible temptation to steal ideas. Honest the reviewers may be but people do fall for such things, even if on rare occasions.

Peer reviewers could potentially slow down the publication of a paper to enable them to get their paper out first. However, reviewers are given a deadline to submit their review.

If a fraudster makes up data carefully, detection of it by peer reviewers may become very difficult. After publication, if a paper is found to be fraudulent or plagiarized, or researchers realize they made a mistake in their calculations that invalidates the paper, the journal publishes a retraction which appears alongside the paper online. These can be tracked on Retraction Watch. If editors are concerned about the validity of a paper and there is an investigation underway, they will publish an expression of concern.

A peer reviewer should be guided by professionals [2]. Peer review may have its limitations, but it is also a remarkable process which relies on the trust and co-operation of the scientific community.

3a.2.2.1 Predatory journals and zombie papers

Universities insist that faculty publish scholarly research, and the more papers the better. Schools they teach at rely on these publications to bolster their reputations. But many such articles appear in 'journals' that will publish anything, after charging publication fees of up to a few hundreds of dollars per paper. These journals are called predatory journals, because they dupe the well-meaning academics into submitting a paper to them. Predatory publishing is becoming an organized industry. Predatory journals do not seriously review the submitted papers they publish. They regularly blast emails to academics, inviting them to publish. The academic system bears much of the blame for the rise of predatory journals, demanding publications even from teachers at places without real resources for research and where they may have little time apart from teaching.

Fraudulent manuscripts can be submitted to multiple journals at the same time: so even if an editor rejects it during peer review, they might see it published elsewhere.

Quantitative metrics are increasingly dominating decision-making in faculty hiring, promotion, and tenure. A hypercompetitive environment increases the likelihood and frequency of unethical behavior during the conduct of scientific research.

A reproducibility crisis is a situation where many scientific studies cannot be reproduced. Inappropriate practices of science, such as selective reporting of positive results, have been suggested as a cause of irreproducibility. Lack of raw data or data fabrication is another possible cause of irreproducibility. Authors should publicize raw data in a public database or journal site upon the publication of the paper to increase reproducibility of the published results and to increase public trust in science.

3a.2.2.2 Dual publication fraud
Simultaneously submitting two manuscripts that report exactly the same study conducted on the same cohort but in different formats, without informing the journals of the dual submission, indicates a fraud—a deliberate attempt on the part of the authors to conceal their behavior. Dual publication must be combated vigorously. Sanctions available to editors include the retraction of the article, a decision to ban the authors temporarily from publishing their work in the journals, a notice sent to editors and publishers of other journals and a letter of censure sent to the authors. Each editorial board must choose from these tools after obtaining advice from its ethical committee.

3a.2.2.3 Retractions
Across the research world, there is a growing suspicion about sloppiness and outright misconduct in the scientific literature. The number of retractions of research papers has been rising, of late. Increasingly, fraud-busters are starting to hunt down manipulated images in published papers and flag them widely. Not many image-checking services have the capacity to rapidly screen a high volume of papers.

If a retraction is the result of an accident or honest error, it should not be a blot on an otherwise respectable publication record. Scientists and journal editors who have retracted papers say that the process can be handled productively, whether the errors are from contamination, a cell-line mix-up, or statistical analyses gone awry. Above all, they say, transparency is the key.

3a.2.3 Open-access publishing

Free access to research findings is good for science and society. However, open-access is clearly not freely open to the scholars who are required to pay exorbitant fees to publish their results, often out of their own pockets. Graduate students who wish to publish open-access articles in the journals of their choice might spend a large fraction of their annual income to do so, unless they have large grants to cover the fees.

The following precautions should be taken before attempting to go for open-access publishing:

(i) Check that the publisher provides full, verifiable contact information, including address, on the journal site. Be cautious of those that provide only web contact.

(ii) Check that the journal prominently displays its policy for author fees.

(iii) Be wary of email invitations to submit to journals or to become editorial board members.

3a.2.4 Publishing a monograph

Hiring institutions are increasingly likely to expect fresh PhD's to have a book under contract. A book-length monograph remains the gold standard for tenure and promotion. The typical advice that doctoral students get from their faculty mentors: write a good dissertation and immediately get it published, so you can get a job. But this advice ignores that the primary function of the students is to get the PhD.

3a.3 Writing of a CV and a résumé

3a.3.1 A curriculum vitae

A curriculum vitae (CV) is a summary of the relevant information about your education and work experience and is usually required for positions within an academic setting (i.e., research, teaching assistant, lab manager). A CV is a record of your scholarly life. Drafting a curriculum vitae for a faculty job market is an important task. Your CV is essentially a running list of your accomplishments. Even if you are not applying for faculty positions, it's useful to have an up-to-date CV on hand—to gauge where to concentrate your efforts to be ready if an opportunity arises.

CVs tend to be longer than résumés since they can include categories such as publications, and lectures. You should design the format of your CV by placing the most relevant categories first. Organize your CV such that your accomplishments can be seen at first glance. Unlike a résumé or a cover letter, a CV is intended to be comprehensive, and you can list in it everything you've done in graduate school and after that.

While a CV is not the only indicator of whether a candidate has the right mix of qualifications, it is the one that enumerates them most comprehensively. For instance, has the candidate taught a course that was expected of them? How many of the applicant's grant proposals are still under review?, etc. A CV lists all this and is a window to what is of most importance to the candidate.

Every CV is unique, because everyone has a different path through graduate school.

Your CV is meant to persuade readers to hire you for the job, give you the grant, or award you the fellowship. You have to select the aspects of yourself that will sell you to your specific audience. People read your CV because they have something you want. When you prepare the document for them, keep your specific goal in mind and customize accordingly. If you're sending your CV as part of an application for a grant, you probably don't need to include your campus service work.

Most readers don't read a CV carefully. They skim it. Use formatting thoughtfully to create emphasis, considering the movement of a skimming reader's fast-moving eye. Format to catch and direct the reader's attention, using perhaps boldface or underlining selectively. Your CV should include only those things that you want to talk about in the job interview. Imagine how you'd like the conversation to go, and organize your CV to make that outcome, likely.

Being able to explain one's research is a fundamental part of being a scientist. There are skill-sets that you're going to need anyway if you're writing for a grant. CVs can be more effective if they allowed room for narratives—brief statements that tell a story about a scientist, their accomplishments or their impact.

An **Impact CV** is distinct from a standard CV. The latter focusses on one's career and covers funding, publications, and teaching and mentorship roles. An impact CV by contrast lists evidence that has been used to build stories for funders, for audiences, for promotion/award committees, etc.

Additional skills that relate to your teaching and research should be included in your CV. These might include any coding languages you know, video or audio editing, qualitative or quantitative data-management systems, experience with scientific equipment and techniques, and software programs prevalent in your discipline.

List any memberships you have in scholarly or professional organizations relevant to your field(s). Such memberships demonstrate that you are an active and interested participant in your discipline.

Remember that CVs are always a work in progress. PhD holders should keep graduate-school teaching (both TA and instructor positions) and service to the graduate-school department and university entered on their CV all the way up to the point they get tenure. Thereafter you can start to trim.

3a.3.2 Résumés

Recruiters spend 6 s on an average deciding whether they want to keep looking at your résumé. Résumés are typically more compact than CVs. A résumé is a compendium of your competencies, accomplishments, and future capabilities. The purpose of a résumé is to motivate an employer to interview you. It should work in unison with a cover letter to emphasize your strengths and document your skills.

Résumés are appropriate to submit for most positions in fields like business and public service. A résumé must have a clear, concise, organized, and professional appearance. It should be easy to scan and key information should stand out. A résumé should be just one page—a general rule to follow is no more than one page for every ten years of experience.

Streamline the top half of your résumé. Recruiters spend a few seconds scanning the top of your résumé when they open it, and then decide whether or not they will invest any more time in reading it. Hence, shift the target employer's industry-specific skills to the top of résumé. Add a core-skills summary and job-specific attributes under your profile.

It is advised to remove extraneous details from your résumé. In certain fields (marketing analytics), experience is only relevant if it is recent. Remove details of your experience, if the position doesn't require as much experience as you have. Retain minimum details on prizes won, internships undergone or fellowships received, and only if they are relevant.

An entry-level role requires a unique set of skills and qualities. Show pointedly in your **entry-level résumé** your capabilities related to core requirements of the job applied for. Entry-level candidates must have a desire to learn, a strong work ethic, and aligned interests, as much as hard skills and relevant experience. A résumé of an entry-level candidate must show enthusiasm and initiative. Pointing out any extra online courses or training attended after your degree would strengthen your commitment to the field.

To handle the huge numbers of résumés received online, human resources departments employ artificial intelligence systems to pluck out the candidates deemed to be good fits. So, while applying is easy, your résumé has every chance of being screened out rather than ending up in front of a recruiter. Tools evaluate résumés by finding keywords related to categories like skills, experience, and education, and weigh them according to the job requirements. Making it through the automated screening needs tailoring your résumé, not just the cover letter, to each job you are applying for. Include in your résumé the keywords, which are the same, or similar to, those that the job posting uses for the knowledge, skills, experience, and duties involved. Use the most relevant keywords in your most recent job listed. Microsoft Word is best to use, but avoid making columns, so that your résumé is readable by AI tools.

Entry-level candidates applying for a professional job are often checked for their basic skills and competency by taking them through one or more tests. Most test takers are then asked to record a video interview. The information from the résumé, the tests and a transcript of the interview is reviewed by artificial intelligence software which screens out the majority of candidates, retaining only around a fifth of them to speak to a recruiter.

For better success, the job seekers should update their LinkedIn profile, and obtain recommendations from managers and colleagues. Their public social media accounts should include digital 'bits' of information highlighting their skills, experience, and interests. Candidates should also seek help from people in their target companies for referrals, to increase their chance of being hired.

The résumé of a researcher ought to be a narrative-based document focused on their four key contributions made, respectively, towards: generation of knowledge; development of individuals; the wider research community; and, to the society, at large.

For some traditional employers like banking and finance, a crisp résumé is important. Look closely at the job description and highlight keywords and skills the company is looking for in that role. Make sure that, if it's relevant and applicable, you have highlighted similar skills or even some of the same keywords on your résumé.

3a.4 Writing of thesis/dissertation

A PhD thesis should generally have the following sections:
1. Title of thesis
2. Summary of work done, highlighting major achievements
3. Introduction to the thesis topic, including motivation
4. Methodology of data collection
5. Analysis of data, drawing inferences of publication value
6. Discussions of the whole thesis work, main accomplishments
7. Conclusions
8. Papers published based on the thesis work.

3a.4.1 Why writing of a dissertation becomes so tough for some candidates

Some students have difficulty in writing their dissertations, possibly because they surround themselves with friends in the lab who are not actively writing a dissertation at that time, OR, are spreading negativity, grumbling about their advisers, or the terrible state of the job market, etc. What you need is a positive community that supports you through the ups and downs of writing a dissertation and celebrates your every success. The finished thesis has no connection with the personal travails you had during your doctoral work.

To get over 'writer's block' while writing your thesis, remind yourself that each doctoral student, including you, has it in them to write their own research work in a professional way. Within their PhD work, the student sets out to tackle some questions, the answers to which either do not exist, or need to be found unequivocally. So, the candidate has to list in their dissertation, what unanswered questions they plan to answer, why those questions are important, why the reader should care about them, and what their plan is to tackle them.

Stick to a writing schedule, and keep your adviser informed about the progress you are making in your thesis-writing. Stay positive and get over the factors that hold you back from writing your thesis.

It could be a lack of concentration; self-doubts about the strength of an argument being made by you during writing; or, getting dejected by the quality of the text written by you, when compared with a masterly finished product, i.e. a thesis written by a past colleague, who finished their PhD degree like a shining star. To get over all that, de-clutter your desk, which will have a refreshing influence on you, before you resume writing. In between writing spells, do go for long walks, alone. It builds creativity.

While writing your thesis, keep reminding yourself of a few things for developing a thesis that you can look back with pride. First, to maintain a good flow, avoid repetition of statements, unless absolutely essential. Remove sentences which are not needed, and dilute your text for no reason. Second, yours being a STEM doctoral thesis, it should have no statements which are equivocal, i.e. conveying more than one meaning. Having equivocal statements or paragraphs in the text of your thesis can make the final conclusions made from your long efforts doing thesis work unimportant or show them in poor light. This could have an adverse effect on your employability, too. Lastly, throughout the thesis, keep focus on the main goal of the thesis as expressed in the title and in its initial summary. Of course, keep showing the rough draft of your manuscript to a friendly reviewer for comments, which you can use to make mid-course alterations for a smooth and focussed text.

The best dissertation (thesis) is the one where the same critical attention is devoted by the candidate to all the chapters, and to each page of it. Don't forget, students yet to come may read your thesis for guidance on writing their own thesis.

Most reviewers/referees/examiners are busy people, and therefore, will not have the patience to read things beyond a page or two, in a serious manner, from which they would make their assessment. Hence, the introduction section may not get a serious reading by experts. So what is it that should be aimed at the reviewers? It is the summary or the abstract of the thesis. Often students think that the summary and the conclusions parts of the dissertation are the same. No. The summary is best placed before the introduction chapter and should be a maximum of 2–3 pages, and should highlight the goals that were set at the beginning, the overall course of the work that was undertaken, and concise claims to new results obtained, and achievements made.

It is suggested that the introduction section of a thesis is best written when the draft of the rest of the thesis has been written. The candidate should rather begin with writing the main body of the thesis, i.e. the middle three or four sections which deal with the original results obtained by them, and discussion of them. In fact, it is convenient to write the results and discussion sections first, since several parts of them may have possibly been written up by the candidate in association with their guide/collaborating co-authors while preparing the papers already published by them on the results obtained during the thesis work, or manuscripts sent to journals or communicated for presentation at conferences. The research guide would, of course, ask the candidate to start writing a thesis only after making sure that the candidate has obtained adequate and new results.

Coming just before the results and discussion sections but after the introduction, should be a section describing details of the experimental techniques and the apparatus used to obtain the data, and how the samples used were synthesized, or how their thin films or single crystals if used, were grown. This chapter is best written after writing results and discussions. Thereafter, go for introduction, summary (abstract), and conclusions.

The conclusions are best written at the end, as a small section of just about a page. The conclusions are different from the summary, because they comprise a quick narrative of the high points of the results and achievements as new additions to the knowledge-base of the topic chosen, ending with some suggestions for the future work, particularly for the benefit of the new students who would read the dissertation.

To sum up, the best way to write a dissertation would be to first carry out all the work under the project, then write all other sections in sequence, except the summary, the introduction, and conclusions. Among these three, the summary should be written first, with the focus of grabbing the attention of the reviewer, avoiding superfluous phrases or arguments. The conclusions can be written next, keeping in mind the points made above. That makes the introduction the last section to be written. The candidate will do a good job of it now, because the rest of the dissertation is already available as a draft text, and the introduction has to be a nice window to it, detailing what was the need for the work that was done under each of the other sections, and what approach the candidate took for that, and how one thing led to another.

It must be noted that writing of the 'dissertation' has to be a candidate's own individual effort. Each dissertation generally includes a statement of 'originality' of the work done and reported by the candidate under it. The research publications ensuing from a dissertation work also have to make a 'conflict of interest' statement.

3a.4.2 Adapting a dissertation for publication, and papers arising from PhD work

Dissertations are digitized nowadays, but it is still quite likely that your work may never be downloaded. Have concrete plans to publish your dissertation, so that your hard work will have more than a few readers. To convert a dissertation into an article, you will have to be brief and present only relevant details.

A key part of the graduate research environment is to disseminate your work through scientific publications. Write papers as you go along, remembering that it is in your best interest to get as many high-quality publications as possible from your work, especially if you want to go into academia after getting your PhD. Each paper should tell a complete story of value to other researchers, but while publishing them, keep in mind the ethical guidelines for publishing.

3a.5 Composing an application for a research grant

To secure a research grant, you need to have an innovative idea that stands out. The main steps are working out the budget and manpower details, and preparing the draft of a project proposal (further details given in next section 3a.5.1.) to submit. Keep ready with a presentation to be made before the funding agency for their approval, when they call you for it. A well-designed presentation can be a great way to structure a compelling grant application.

Send a one-page sketch of the abstract and aims of your project to your mentor and co-investigators. Seek external reviews prior to submission. Putting your proposal through a mock review panel can vastly increase your chances of funding.

Early-career grant application should be a focused methodological plan directly tied to your specific aims. Getting even a small grant award will show that you can succeed in grant-hunting. Funding agencies may post a list of prior and current grant reviewers and their affiliations online. Review the list to check if their expertise overlaps with the aims and methodology of your study. It would be a high-risk proposition to write a grant for a foundation that has never funded an application in your area of expertise before.

3a.5.1 Project proposal

Planning a research project involves defining your research question, and reviewing previous research on your question before formulating your own research plan. Your research proposal must indicate that, if accepted, it will have something novel to contribute. Only pose research questions which can be answered under the prevailing constraints of time and money. State how the approaches developed in this project will help in other areas of research. In addition, your specific aims should bring out a clear alignment between your proposed study objective and the agency's mission statement. State at the end, the impact of your work, if funded.

The proposal should provide the following details. Title of project; summary; why do it at all; aim of work; introduction of topic; current status of the work done; equipment/consumables/space/students needed to complete the work; what has already been achieved by you in your lab; plan of work semester-wise; what your institution will contribute for the project; deliverables anticipated at the completion; list benefits to be accrued – to society, to subject, to students, and to Institutes; mention uniqueness of the proposal; provide track record of the principal investigator (PI) with CV and list of projects completed already.

Project proposals seeking grants to conduct research are subject to the 'peer review' process. Funding bodies seek expert advice on a scientist's proposal to select which projects to fund.

3a.6 Writing applications for jobs—a cover letter

A cover letter is a letter written by you to introduce yourself to potential employers, showing how your skills meet their job needs. A well-written cover letter connects your qualifications to a specific job with a prospective employer. A cover letter must be written fresh, specifically addressing the employer's job description, each time. A cover letter should be concise, just a page, if it is for an entry-level position, with relatively short paragraphs, and can have an introduction, body, and conclusion. While concluding, reiterate how your education, training, skills and experience fit the job, and that you look forward to a discussion on this further.

Typically, a more advanced position will require a lengthier cover letter. In all cases, however, remember that the readers of your cover letter are busy people. Therefore, keep your letters from extending beyond two pages, unless the job advertisement specifically requests more details about your recent research or work experience.

Never send out the same cover letter each time. A tailored cover letter for each faculty job will make you appear both more prepared for the job and more enthusiastic about it. Not tailoring your letter implies you have not clarified why you want to join that institution. It should highlight why you think you fit the position well and why you think you have the skills to succeed in this job.

The best thing before writing the cover letter is to study the ad and look for details, such as: area of specialization, type of institution (public, private, research- or teaching-focused, etc), mission of the institution, teaching the candidate may be expected to do, teaching load, research expectations, and qualifications. Be clear whether your letter focuses on a teaching job, a research job, or both combined.

Do not use someone else's cover letter as a template or use ChatGPT to generate the cover letter. These can become reasons for rejection right away.

In the opening of the job letter (cover letter), introduce yourself and your purpose for writing. Identify the position you are seeking by name and state how you learned of the position. Establish that you have at least the minimum requirements for the job by listing your specific academic degree and any immediately relevant work experience. Just after that mention how your skills directly relate to the job requirements. Support your case by stating any internship or specialized courses that you have done, directly relevant to the needs of the job mentioned in the ad. At the end, invite the reader to view the attached résumé. Mention that you are available to provide any further information, or directly for an interview.

3a.7 The letters of recommendation

Ads for academia jobs routinely request letters of recommendation. Most new PhDs will be applying for several dozen academic positions during their first job search. Each one of those applications requires references, say a few per candidate. But a good, strong reference letter has to be tailored to each position, making the recommender invest time in so many letters.

If you are a new PhD, your dissertation chair/adviser must write a letter in support of your candidature. The absence of that letter would be a cause for concern. If you did your PhD a while ago, then your current department chair or group leader should write a letter. Give each writer a specific topic so that, individually, they don't all cover the same ground.

Do not write your own recommendation letters and have someone sign them. Request your writers to tailor each letter to the specific position. A writer who does this communicates deep support of your potential.

Letters of recommendation become the make-or-break of an application. Search committee members eliminate applicants who don't have the required number of letters or whose letters are too short, or lacking in valuable information. Of the remaining applications, they read every recommendation letter carefully before arriving at a consensus.

3a.8 Research statement

A research statement is a component of academic job applications. It is a summary of your research accomplishments, current work, future directions, and potential of your work. The research statement should be technical, but at the same time intelligible to all members of the department, including those outside your sub-discipline. The best research statements present a readable, compelling, and realistic research agenda that fits well with the needs, facilities, and goals of the department that you are applying for.

Why is a research statement needed? It conveys to search committees the basics of your professional identity and plots the path of your scholarly journey. It communicates an assurance that your research will follow logically from what you have done, essentially providing a context for your research interests. It combines your achievements and current work with the proposal for upcoming research.

Finally, the research statement helps the hiring committees to assess your capabilities, viz. your areas of specialty and expertise, potential to get funding, academic strengths and abilities, compatibility with the department, and your ability to think and communicate.

What does a research statement do? It serves as your introduction to a search committee, which would comprise of scientists from your field, as well as outside your field, and acquaints them with your research. For it to be eminently readable, make it short, say just one or two pages.

How to write a research statement? First, write about the importance of the main theme underlying the subject area of your research, and what specific skills you use to tackle the problem. Give examples of some specific problems you have already solved, to build credibility and write about what you do. Include a discussion on the future direction of your research. If you believe your research could lead to answers for big important questions, then do state that! Finally write a paragraph about what exactly your research is and what its aim is. It should be just a summary, keeping the details for the job talk, i.e. presentation before the search committee and interview.

Your aim should be to convince the search committee that with your knowledge you are the right person to carry out the research. Include preliminary results and discuss how you plan to build on them into an exciting research output. Using a summary of your research, describe how the faculty may become partners in it. List major findings, outcomes, and implications. Describe both current and future research. Communicate that your research will take off from your recent work and that it will be unique and innovative, and will consume only modest funds. Discuss briefly the major research problems you want to focus on, and why they are relevant.

Prepare a longer version (5–15 pages) for use during your interview. In the campus interview you should describe your research plans and budget for equipment in detail, and list how many graduate assistants you will need in your laboratory to start up the research. Identify potential funding sources, including any external funding that you would be able to get for research. Mention past funding in support of your funding plans.

3a.9 Preparing a career portfolio

A career portfolio gives you a unique professional identity that evolves with you. It gives you ownership of your career, because unlike a job that someone else gives you, a portfolio can't simply be taken away. It is yours forever.

It reflects your professional potential and projects your unique combination of skills, experiences, and talents that can be mixed, matched, and blended in different ways. Include in it any role or activity in which you've created value and served others: freelance roles, volunteering, community service, side hustles, passion projects, hobbies, exchanges, etc.

As time goes by, the value of your portfolio will increase by your ability to cross-pollinate, i.e. to combine and weave together skills from your different experiences in order to gain new insights, tackle new problems, diversify income sources, and serve in new ways. Your career portfolio is naturally aligned with your lifelong learning and is meant to help you expand your professional community and access to leadership opportunities.

Those who build a career portfolio will be more prepared to pitch themselves for new opportunities, as they will be able to make quick creative connections between their various skills. Employers are hungry to hire talent with non-traditional backgrounds, but they often need help. Your portfolio narrative is the link, the bridge, and the story you tell to make connections between the skills people are hiring for and the skills you have developed through the breadth of your experience.

3a.10 Becoming a PI

The biggest hurdle in an academic career is rising from postdoc level to a PI, which means moving from doing someone else's research to getting other people to do yours. Both the jobs happen to be in the same environment, but it is not an easy transition. A postdoc has a good chance of becoming a PI (eventually) if they are creative with new viable ideas in research, have a good track record of publications, can write science well, make their research output like a brand, and build a network of supportive people. Remember that rapid expansion in PhD programs without much addition of new faculty posts in a department can create a situation that a very small number of fresh doctorates will become PIs.

3b.1 Public speaking with confidence

We all often wonder what brings that 'super'-confidence in an expert speaker in science subjects. Is it mastery over subject; or clarity of thought; or conviction about the topic; or inspiration drawn from another iconic-speaker; or rehearsals of ideas with colleagues/self; or ability to alter the presentation, on the spot, to suit the audience?

3b.1.1 Public speaking skills for doctoral students in STEM

One of the most important skills that STEM students need to develop during their graduate career is the ability to communicate their work to a wide array of audiences. That ability enables STEM students to speak effectively about their research with scientific peers, as well as to walk into an informal setting to tell an audience of all ages about the same exciting research. STEM students encounter these audiences in the course of their careers.

Evidence suggests that skilfully crafted communications influence not only the public but also scientific colleagues. It is no accident that the top scientists in any given field are also among the field's best communicators.

Many graduate students fail to talk in simple language when describing their research. By the time they finish their doctoral work, they need to become expert storytellers who know how to craft tales that accurately reflect the science hidden in the narrative that readers/viewers learn without realizing it.

A good presentation is delivered slowly. It has expression (pauses, rises, falls, and stresses), and includes a story, metaphors, or emotional elements. It must give concrete examples, and tell even the experienced audience at least one thing new about the topic that they didn't expect to hear.

STEM researchers, at all levels need to develop presentation skills to help them in their job prospectus and career-growth. Time spent on improving presentations should not be taken as detraction. On the contrary, this directly helps in career development, such as in writing compelling grant proposals. Designing a good presentation makes a researcher organize data into themes; it's an effective way to consider your research in its entirety. A comprehensible presentation should elicit probing questions from the audience, which in turn might generate further research ideas.

There is at times a disconnect between researchers and their audiences. Academics, in general, don't think about the public. Their intended audience is always their peers whom they wish to impress to get tenure. Academic narrative is often loaded with professional jargon and complex syntax [3]. So much so that even a PhD holder can't understand a fellow PhD's work unless they are from the same discipline.

Simplicity should be the hallmark of scientists during public speaking. Communicating effectively with subject experts automatically improves the presentations made to larger non-expert audiences such as the media or the individual donors. Senior scientists should provide opportunities and training to young researchers working with them to making effective presentations.

Try attending a plenary lecture, even if it is not on a topic of your choice. It will give you the chance to see the work of someone who is top-notch. Listen not only for what they say, but how you can pick up a lot about crafting a talk, speaking to a large audience, and answering questions.

And it's not just the spoken and written word, body language is important in communicating successfully. Pay greater attention to how people are speaking with their body language than what they're actually saying. Find innovative ways to make an emotional connection with the public through body language and story-telling. For body language, learn from professional actors who communicate with the public so effectively, with their non-verbal cues.

Your presentation should be like telling a story comprising a clear introduction describing why you are doing this work, and why it's important; a middle section, and an end in the form of conclusions and suggestions for future work. The concept of telling a story also applies to writing papers and grant proposals.

When the allotted time is short, make sure you convey only the really important concepts, and skip details. When you are finished, you want the audience to remember the key points of your work, and not the petty details.

3b.1.2 Useful hints for successful public speaking

The conference talks we remember are the ones where the speaker made eye contact with audience, spoke slowly and confidently and made at least one point well.

Plan how you can communicate ideas you are quite familiar with to a non-specialist group. Practice, practice, practice, alone, in front of friends, and in front of small audiences.

The slides in a presentation should be few in number, say 15 for a 25 minute presentation. Too many slides dilute your message. The audience will also lose interest if slides are totally made of text. To hold their attention provide pictures or graphs to describe your data.

Most effective science communication really comes down to just telling a good story. Knowing your audience and your communication objective is crucial, but if you can turn your message into a story, it has a much better chance of being accepted. Stories increase people's likelihood of remembering information, and are more compelling and convincing than even facts.

Use of a narrative helps to convey scientific findings in a coherent manner that can help the audience better understand and remember complex processes that are otherwise difficult to explain. A broader narrative captivates attention by promoting identification with the story and eliciting deeper emotional reactions, keeping the audience anticipating developments and conclusions. Scientific narratives explain science to common people who can then use their improved understanding to make better decisions for themselves, or reconsider long-held beliefs that may be inaccurate, and develop understanding that will serve as a framework for new information.

Use a conversational tone in which you are explaining things to the audience rather than giving them a speech. During your talk, smile. Smiling suggests to the audience that you are happy to be speaking to them and that you care about communicating with them, and that you are having fun explaining science to them.

Structuring a speech is important so that all goes as planned even if you are asked to complete it in a time shorter than originally allotted. A well-structured speech starts with a summary—the gist of the presentation you will be making. The middle of the speech presents the information supporting the short summary presented. It also has facts to support the main theme of the work, and stories or anecdotes in support of your work. At the end it gives a take-home message to the audience, in line with the short summary presented in the beginning.

To capture the attention of the audience, a speaker can use what is called an *attention getter*, which could be a short anecdote, a statement or a question.

Do not exceed the time scheduled for your talk. If you have been given a 20 minute slot at a conference, plan to talk for not more that 15–16 minutes. A good presentation needs to be accurately timed and the speaker needs to be sufficiently confident with the material that even if the power fails or a wrong button is hit and the slides disappear, he can still keep going.

Stand up while speaking, to let people in the back rows see your face and hear you better.

Move around while speaking. Hand gestures are good, use them to show your enthusiasm for your topic. Vary the pitch of your voice, monotones are boring. Speak loudly and clearly. Continue to face the audience when you speak, even when you are using visual aids, like slides.

Make eye contact with your audience. Locate a few friendly faces at different places in the hall and speak directly to them, switching often from one to another. But be careful not to ignore the rest of the audience.

Summarize your talk at the beginning and again at the end, to help them grasp and remember your main points. In other words, at the outset, tell your audience what you're going to tell them, and before closing tell them what you told them.

Ask the organizer to remind you when you are left with just 5 minutes, so that you can still re-organize and end logically.

Lastly, try to emulate excellent speakers. Model your talks on theirs.

3b.1.3 Fear of speaking in public

You may fear public speaking, even the idea of making a presentation before an audience or at a meeting makes you nervous. Public speaking is a skill. It takes time and practice to build this skill and comfort level while speaking to a big or small audience.

Lacking confidence can create stress. If you are stressed, your audience will see it. Training and practice will help reduce stress, build confidence and skills, and increase your comfort level with public speaking.

Most people get nervous before a speech, presentation, or important meeting. Many people believe that a certain amount of nervousness can keep you focused and enhance your performance. It is good to analyze whether the fear is about standing in front of a large group of people, whose eyes are on you. Or it is the fear of going blank, freezing up before starting to speak? Identify the specific cause freaking you out before public speaking. Having understood the cause, work on an alternative to beat the cause.

All speakers have fear—they just learn to *manage* it. Some amount of stage fright before and during public speaking is normal and it shows that you care and are keen on delivering an effective presentation. One way to get over the nervousness is to try to meet your audience before your talk, and even visit the room or auditorium where you are going to talk, after, of course, preparing your talk well.

When receiving a question, do not rush into answering it. First, listen to the entire question to make sure you understand it; then take time to construct a to-the-point answer. Look at answering a tough question as an opportunity to show your hold on the subject.

When nothing works: audio, laptop, mic, etc, keep cool. Have a plan B. Speak loudly without mic.

Doctoral students in STEM subjects should be aware of the reasons why they should be naturally good communicators. One is that most scientists start with the engaging quality of enthusiasm that they maintain during their PhD research work. With enthusiasm on their side, they can exploit their natural gifts of clarity.

Scientists can think clearly, can usually write and speak clearly, because they have the capacity to expresses their thoughts clearly. In fact, during doctoral work, a PhD student evolves into a keen observer who can make distinction between things which are different only in a subtle manner.

Difficulty with public speaking has less to do with the fear of speaking itself and more with how we constantly reinforce that fear. The remedy is to face it, get up to talk, even if a voice in your head reminds you of your fear. Conquer stage fright by speaking often, and speaking in front of a variety of audiences, and getting used to it. Learn to engage your audience by structuring your presentation well and improvising it, whenever needed.

Look at your slides before your talk. Is the spelling correct? Is the sequence of details in the slides the same as the sequence of arguments that you will be making while speaking?

Great speakers know their lecture material well, thanks to multiple rounds of rehearsing. The most moving speakers try to relate their material to happenings in their own lives. They also use anecdotes make the message more relatable.

The message your body language sends is just as important as your verbal message. Your volume, tone of voice, eye contact, facial expression, can all work together to enhance your presentation. Talking slowly and inserting the occasional pause does convey a sense of command. That lets audiences know they are in capable hands.

The best speech is one where people leave with something new and interesting to ponder. But that requires careful planning by the speaker, contemplation, as well as a focus on the topic, and ample rehearsals.

Review a recording of your own performance on stage while communicating on a research topic as a doctoral student. Are you looking at your slides too often, and turning your back to the audience? Are you avoiding making eye contact with the audience? You may have to attend to these actions, to deliver a memorable presentation. Finally, the best advice to deliver a really good talk is to stay calm and confident, put some passion in your voice, and see that each of your words are heard by the audience, and make eye contact with listeners to engage them and support your words with motions of your hands and body.

3b.2 Making a conference presentation—oral or poster

As you grow in your career with a STEM PhD, you will be called upon to deliver invited talks, during which you are expected to give a comprehensive overview of the field, apart from discussing your own work. As you grow further you will be invited to deliver a 'Keynote' address, mostly with enhanced time slots of say 45 minutes, a prestigious thing to do, at a conference. Finally, you may qualify, depending on your academic standing, for the ultimate honor—a 'plenary' talk, which means there is no parallel talk at the conference during that slot.

Other than conferences, research institutions and university departments, hold seminar programs, in which speakers are invited to deliver lectures lasting

45 minutes to an hour on a theme, which includes a question–answer session and a general discussion.

The difference between a seminar and a colloquium is that a seminar is usually meant to be delivered to a specific interest group, whereas colloquia deal with a more wide-ranging topic and are, therefore, intended for a broader audience.

3b.2.1 Oral presentation

First of all, check the allotted time you have for your presentation and make sure your talk fits into it. Be aware that a 20 minute presentation, will give you just enough time to make an introduction, discuss one or two main issues, repeat some important points to help the audience follow your talk and remember it, and finish on a firm conclusion.

Rehearse the presentation. Understand the background information and data that you are presenting, in case questions arise. This also will diminish anxiety and facilitate a relaxed interaction with the audience in a professional and confident manner.

Rehearse your talk, first with a local audience, who should give useful feedback to you in order that you can make distinct improvements in your presentation to gain confidence. Give a concise introduction providing the current status of the field, and point out where your work comes in. State why your field of research has potential, and what attracted you to it and what contributions you have made.

Don't make slides crowded with a lot of text. The audience is trying to read and listen at the same time, so too much to read in every slide will distract them, and make the task of processing the oral information less effective. Use graphs or illustrations to support your story, because that's what the audience really likes, and it also helps them process and remember the information.

A major distraction for the audience comes when the speaker looks repeatedly toward the projected slides during the presentation. No wonder his eye contact with the audience is lost, and the audience loses interest.

Whether speaking live in front of an audience or presenting virtually, plan for one and a half minutes per slide. Assuming the listeners' limited attention span, knowledge of the topic, text readability, etc it is advised not to put more than five bullets point and more than five words per bullet point, in a slide.

Delivering effective oral presentations involves three components: what you say, how you say it with your voice, and everything the audience can see about you and your talk. For all three components, maximize the signal-to-noise ratio: amplify what helps, filter out what hurts. Speak extempore. Instead of using fillers like *um*, simply pause. You can get up to two or three seconds of thinking time without the audience even noticing it. Vary the tone, rate, and volume of your voice as a function of the meaning, and importance of what you are saying. When you make a gesture, make it large. Engage the audience by looking them straight in the eyes.

While answering questions after your talk, let the questioner finish asking the question before you answer. Don't panic, rest assured, you know more about the topic than the questioner. It is OK to pause for 5 seconds before answering a hard question. If you don't know, say so. Don't bluff. Have backup slides for key questions.

Oral presentations at a conference or internal seminar must convince the audience that the research being reported by you is important. Hence, it must emphasize the *motivation* for the work and the *outcome* of it, with enough evidence to establish the validity of this outcome. Project professionalism through well-chosen words, steady delivery, and a professional authenticity. Use your voice to maximize the impact of your main points.

A presentation at a conference aims to present recent advances, whereas a presentation at a PhD defense aims to inform the audience the research work done during the doctoral study, and its rationale. Oral presentations are for convincing. That is why oral presentations normally have a questions and answers section at the end. It is a good idea to allow enough time for questions and answers, which can be done by saving time by being selective during the main presentation.

3b.2.2 Poster presentation

A poster presentation is a form of oral communication at a conference. Design your poster like a set of slides rather than a paper. Select a few messages. State them verbally and illustrate them visually. Organize these messages into a layout on your poster. Relegate details to a handout. Structure your poster for maximum attention by viewers. In the main clause, put what is new/interesting to the majority of your audience. Visual material is crucial for non-specialists. Drawings are best for conceptual explanations because they focus on the essential idea, doing away with unnecessary details. Photographs, give a better idea of concepts to non-specialists.

Delegates who attend a poster session at a conference are generally wandering through a room full of posters, full of people, and full of noise. Unless they have decided in advance which posters or presenters to seek out, they will stop at whatever catches their eyes or ears, listening in on explanations given to other people and perhaps asking an occasional question of their own. Therefore, promote your poster before your poster session, make people curious about it, invite them to come to see it. During the poster session, be proactive. Welcome attendees to your poster with a smile. Strike up conversation, manage the flow of questions and visitors. End each conversation by giving a handout.

A poster session is like a bazaar with many poster-holders competing for attention, simultaneously. Among delegates visiting a poster session are professionals well-versed with a given topic, as well as the amateurs, who know very little. To gain maximum attention for your poster, you should organize your poster as if it is telling a logically effective story that you can tell whenever a delegate shows up.

If your visitors like your work they may actually end up reading your papers on the subject and either offering to collaborate or at least citing your results in their own work. Therefore, in addition to the sign-up sheet for requests, you should have with you some reprints of the work you are describing in the poster, together with a lot of business cards/handouts with your contact details on them.

Volunteer to answer questions related to your poster. But, be brief while explaining. With so many posters to see, visitors have only limited time. If they ask focussed questions, it means they need more information, then feel free to go into details. Still be watchful of other people waiting to ask you different questions. They may soon decide to move on. Be ready to give the explanations, even if all over again to new people. Treat all visitors, including the last one, to your poster with the same enthusiasm.

3b.3 Making full use of participation in a conference: learning to network

Professional development is an important component of academic careers. In the weeks after a conference, try to strengthen the connections you made. The first thing to do after a conference is go back through your handwritten notes and look for any bits you've highlighted.

Early-career academics at conferences should be primarily focused on networking. Networking is definitely something you should learn. It is one of the things that is directly responsible for getting a job. So network early and often. Seek opportunities to talk with experts and scholars at a conference. Most of these opportunities are embedded in pre-conference events and in gatherings after the day's sessions are over. It is advisable to build contacts and professional relationships with people you met as participants at a conference.

Doctoral students, and young researchers who just started attending conferences should not hesitate to discuss or make queries. Asking a question when a speaker invites them will build your own confidence for on-the-floor discussions during a session. To get ready with an appropriate question to be asked from a veteran speaker, you should first look at the abstract of his talk in the abstract booklet. Remain alert during the talk, and jot down your query and keep it ready for the question-time. A pertinent question will be welcomed by the speaker. This will help you to follow up with the speaker later via email. You can then build up a dialog with them after returning from the conference, making a reference to the query made by you at the conference.

Introduce yourself at a conference, go to the social events and receptions, and make an effort to get to know people. Try to become a part of the luncheon table/coffee-time discussions between groups of speakers, after introducing yourself. It's these connections you make that really sustain you throughout your career.

After attending a talk, introduce yourself to the speaker. Offer a friendly question. You have already made a connection. During the conference itself, note down the contact details of the conference speakers whom you would like to make contact with after the conference. It is not difficult to find out these details from the internet, too, with a bit of effort. Try to study the recent contributions of a speaker before making contact. Email is best. Make a mention of the conference just attended where you heard them. Write a line or two of their work to show your interest in a particular aspect of their talk/work, and ask for further comments. Try to mention a published report of the work of this speaker, with its title and reference, to underline the seriousness of your interest in their work. Rest assured, most expert speakers, even Nobel laureates, do reply if you have made a query which deserves a clarification.

Keep looking for opportunities when you can meet this expert speaker once again at another conference. If this is not possible, then try and suggest their name to a conference organizer known to you who would invite this speaker. Meeting the same speaker second, third, or fourth time, etc, helps in developing a long-time professional contact with them. And don't forget to send Christmas/New Year/birthday greetings to this speaker on a regular basis.

After receiving a positive response from a speaker at a conference or a fellow-participant at the conference, try to keep the correspondence going, by regularly discussing the further work of the speaker and even presenting your own views, as well as sending records of your own recent work, if relevant. If this dialog sustains then chances are that at least in a few cases it can develop into a professional friendship, which is useful for getting the references of such eminent speakers, when you apply for a position.

Fresh PhDs, and doctoral students have the most to gain from attending a conference. Big conferences are fine, but initially, it is beneficial to attend smaller regional conferences, or meetings that directly involve your research. Write a review of a conference for a journal, or a news outlet. Whether you volunteer to review conference proposals, or help steer the conference, organizers often reciprocate with free or discounted registration.

3b.4 Preparedness for a job in academia/research lab or in industry

Although the progression from graduate student to postdoc to tenure-track to tenured professor appears pretty direct, this progression actually holds for a minority of PhD careers. Graduate programs keep preparing doctoral students for careers as faculty members at universities, despite the well-known paucity of academic jobs and efforts to diversify the options available for doctorate-holders. The result is that most students are neither able to secure faculty positions, nor do they have a clear concept of their suitability for work outside of academia.

To make your dissertation worthwhile, think about the job you might wish to apply for upon obtaining your PhD. If you are considering a faculty job, how will your dissertation make you stand out from 100-plus other applicants seeking the same position? Which are some conferences where you can start sharing your results and expanding your academic connections? If you are seeking a non-academic career, how might your research findings reflect your day-to-day work? What will you say about your dissertation during a job interview?

For academic jobs, be ready to answer the following queries:
1. What is the question that you were seeking to answer in your research work?
2. Explain why your work is so important in just four or five sentences.
3. Cite the limits of previous published work. In that light state the importance of your own proposal.
4. If your project allows, give a simple narrative of your work proposal.

Successful PhD holders, regardless of whether they stay in academia or transition to industry, take initiative. Before submitting your CV, approach the hiring manager through the professional social-networking platform LinkedIn or just an email. Keep your message brief, introduce yourself and attempt to learn more about the role of the job position advertised.

Companies hire people who are critical thinkers, who can separate what is important from what is not important. They need broad competence as opposed to exclusively a high-level of knowledge in a particular scientific niche. They want people who will be able to contribute for the long haul. Your responses to interview questions can leave a good (or bad) impression of your competence, but it will ultimately be determined by the company after analyzing your experiences and success stories.

Most PhD holders are by training proactive having been in an academic setting, and it is natural for them to approach potential collaborators. But, when you are job hunting, you need to take a proactive stance with a more 'commercial' mindset. To get a job you are selling your skills in the job market. One skill of a PhD holder is to be able to analyze complex information quickly and communicate the key message concisely, which is valuable in any field.

3b.4.1 Industry job interview

Come prepared with enough knowledge about the company that is interviewing you. Make sure you know the company's mission statement and values. Consistent eye contact reflects self-confidence, friendliness, and a willingness to engage with your interviewer. Provide the friendly smile and eye contact expected from you by the interviewers.

During a job interview, describe your skills by using the terminology that a potential hiring manager can appreciate. Use anecdotes that demonstrate your qualifications and value.

Your résumé is likely the reason the hiring manager called you. Although you submitted a digital copy with your application, remember to bring 2–3 printed copies of your résumé to an interview. Most hirers will be impressed when they notice that your résumé is customized, highlighting relevant skills and using certain keywords suited to that particular company or position. Write down also the major claims you will make about yourself in a job interview. Then match each claim to a well-developed story. A personalized cover letter can also help you stand out from other applicants for the same job.

Additional experts may be pulled into the interview process at the last minute. Hand them a copy of your résumé, walk them through your career story, and tie your qualifications back to the position at hand. Based on your résumé, develop a narrative that explains how your previous experiences have shaped you into the role at this company. For that re-read the job description before your interview.

The average recruiter will look at your résumé for only 6 seconds, on average, before making a decision. Hence, tailor your resume to each job you apply for. Do choose which of your skills, responsibilities and achievements you select to highlight for a particular job.

Let your résumé be short, preferably not exceeding a single page. At the top, put a 3–4 sentence summary that outlines your work experience and what stage you are at in your career. This should align with the description of the job you're applying for. Then add an attention-grabbing bulleted list of skills.

Lest the interviewers/recruiters think you have been job hopping, put in your résumé (in parentheses) next to the job you had: 'company closed' or 'contract position' or 'downsizing due to Covid-19'. List awards and accomplishments such as efficiency or time management on your résumé. Many employers will compare your résumé to your LinkedIn profile, so the dates and details should match. Recruiters may call you if you have a great résumé, but what is going to get you the job is networking and reaching out to people that are in a position to hire.

Most PhD's looking for work have been working non-stop on their dissertation or fellowship. This work has been your life, and it has shaped your sense of self-worth. But, don't start mentioning the details of your work and discoveries. That's a big mistake, as you might be losing sight of the big picture. In job interviews, mention things only when you are asked to.

Interviewers can ask questions from several topics related to the job being offered. But interview committees are short of time, so they quickly bring the candidate to focus. The best thing is to behave normally, and stay cool and positive, and first listen to a question fully before giving your reply. Any one-upmanship, or hubris while giving replies may be counter-productive

While tackling a difficult question, it is prudent to treat the questions being asked from the perspective of your own qualifications and capabilities. If you can steer the question to your own regime and then answer it by drawing from the best of your experience, it will appeal to the committee. Treat all questions being asked as meaningful while attempting them.

3b.4.2 Facing an academic job interview, online

Postdoc researchers appearing for an interview should prepare by reading published papers from the lab they have applied for. This will help a potential postdoc during the interview, because armed with the knowledge about the work done at the lab, they can convince a group leader better about their seriousness to work in that lab.

Draft concise 3–4 min responses to each of the main anticipated questions on your research, publishing, teaching, and your fit with the department. Rehearse these answers for smoothness.

Be Skype-proficient, keeping your whole head and part of your shoulders visible on screen, and looking level into the camera. Place your laptop on a box if required.

3b.5 Handling tricky questions during a job interview

A few questions often asked during a job interview, which candidates find difficult are listed below, with hints on answers:

Tell us about yourself?
While facing an interview, you can answer this question in two ways. You can either provide with a short 1 minute response, OR you can treat this as an opportunity to impress, opting for a longer, more detailed response. Why not briefly mention your recent achievements in life? In any case, the interviewer has your CV, which gives your career profile. So the best thing is to give a reply that neither shows you off as someone vastly more accomplished than you are, nor seeks a sympathetic attitude towards you.

Any weaknesses you have?
Perhaps, you can say that you have a weakness of not rushing to do things which requires adequate preparation to do a job satisfactorily; you take a bit of extra time to plan it. Such a reply can perhaps shift further attention of the interviewer from this question.

What did you not like in your previous job?
The answer to this can possibly touch an emotional cord. Do not denigrate your previous job conditions even if those were not to your liking. Instead, state that most things were OK in the previous job, but you have been looking for bigger challenges which you believe you may come across in the job that you are being interviewed for.

Describe an example when something didn't go as planned and how you dealt with that?
There is nothing very unusual about things not turning out the way one planned. So, don't be flustered, but come quickly to the focus of this question and recall how you handled a task and could turn that from a failing option to a satisfactory result. Honestly pick up a situation from your own personal experiences, perhaps during your higher education, or maybe at a family function, or at a sports meet, etc. It is ok if you can pick an example that may be out of the box, or even humorous. That will boost your confidence during the rest of the interview.

What salary do you expect?
The ad for the new job already described the scale and pay that goes with it. Don't get into arguments to seek a higher remuneration right away, making you appear greedy, even if unknowingly.

Where do you see yourself in the next five years?
This question is most likely meant to test your ambitions. The safest reply would be to state that after joining this company, you will work hard to learn things fast and contribute to the progress of the company, and would eventually like to rise in the hierarchy by one level in five years.

Do you have any questions for the interview committee?
You can ask a harmless question such as when would you be expected to join, if selected, etc.

All in all, try to share with the interviewers a clear narrative of your experiences, roles, and achievements, possibly in the early part of the interview itself. This will most likely ensure that most of the follow-up questions are asked on your strengths, which in turn raises a good possibility that the interviewer will remember you, accordingly, while deciding who to recruit for the job.

3b.6 Transferable skills for industry jobs

PhDs who don't make it to a faculty job must seek jobs in industry, or in other science related careers like in the offices of patent attorneys, publishing houses, etc. PhD graduates must be armed with transferable skills useful to industry, to gain employment there.

Your dissertation probably won't change the world, but with the skills you gain in writing it, you will be better equipped to make a bigger splash in your field. Academic language is impersonal and cold compared with mainstream forms of communication. Hiring managers need to determine the specific parts that you played in your research. So, when describing your experience, focus on what you personally achieved, even when you worked in a group. Your succinct description of your capabilities, and your awareness of how transferable they were, are key measures that you would be assessed by a hiring manager.

Ensuring that STEM PhDs have material evidence of transferable skills is important for employment in industry. Preferably, during their PhD days, these candidates should acquire transferable skills, such as in data analysis, public engagement, project management or business, economics and finance. The value of such training would be even greater if these skills were to be formally assessed alongside a dissertation rather than seen as optional.

When asked about your experience, the interviewer is trying to determine your relevant skills. You should present a concise and coherent story, not a detailed, chronologically accurate blow-by-blow account of your research. Tailoring your application to highlight only the relevant information will help the interviewer to quickly determine your suitability.

Soft skills such as creativity, problem solving, communication, public speaking, teamwork, punctuality and learning from criticism that you picked up during undergraduate days will serve you well, too.

Corporations spend a lot of money on business and functional skills training, but a great effort is spent towards developing soft skills such as influencing, coaching, listening, feedback, and delegating. When viewed through the lens of how they are applied, soft skills are really interaction skills. The biggest challenges seem to be maintaining self-esteem, clarifying what others are saying, empathy and developing others' ideas. Of all the leadership soft skills, empathy is arguably the most critical. It is the linchpin soft skill. It's the cornerstone of smart leadership. The real competitive advantage comes from one's capacity to create relationships, and in that empathy will count more than experience.

Talent analytics are creeping into the human resources community, enabling chief learning officers and their colleagues to answer key questions, determine what works and what doesn't, and improve talent practices including leadership soft skills development.

3b.6.1 Teaching skills

Strong teaching skills strengthen a PhD scientist's career, whatever direction it may take. Graduate programs pay little attention to teaching scientists to teach. As we train the next generation of scientists, we should help students develop skills as educators. Whether they formally teach or not, scientists need to explain and make science compelling to non-scientists—industrial managers, government policy-makers, patent examiners, the world. We need to adjust our priorities and correct this historic imbalance of only learning how to practice science but not how to teach it. By correcting this imbalance, we will educate an entirely new generation of scientists who offer improved classroom teaching and more accessible public communication about science.

Educators no longer focus on the technical outcomes with STEM, but rather the skills students learn through it. Education planners are designing their STEM curriculum to prepare students for more than just technology and engineering jobs. By integrating technology into lessons regardless of subject matter, educators can teach specialized skill sets and mindsets that can effectively prepare students for the Fourth Industrial revolution.

3b.6.2 Skills from STEM education

STEM education helps students to develop innovative skillsets and mindsets that foster creativity, collaboration, and problem solving abilities. Skillsets refer to the ability to carry out a task to solve a specific problem, like programming, data science skills, and simulation skills. Mindsets refer to how students approach the world around them, the social-emotional skills they use, and how they solve the problems they face, such as through computational thinking.

STEM education not only helps prepare students for jobs in tech, but for a much wider variety of employment. Since we live in a tech-driven world, technology is implemented into aspects of daily life and work. Farmers, lawyers, doctors, film directors, bank tellers, and musicians all use some form of technology to be successful. This could be using simulation technology to defend a client in court, using editing software for a movie, or even analyzing data for a bank. By using tech in STEM education, teachers and IT administrators can best prepare students for a tech-driven economy.

3b.6.3 Skills in science of communication

Just as there is science to be communicated, there is a science of communication. Indeed, there are many ways science is communicated. Communication scientists know something about how messages flow through diverse communications channels, how stakeholders interpret them, and how those processes affect beliefs, attitudes, and behaviors about science and scientists.

Some of these communication scientists are found in disciplinary departments, such as psychology, sociology, and political science. Some are in interdisciplinary ones, such as geography, business, and public policy. Communication departments include scientists who view themselves in both ways.

Physicists can help to develop more realistic models of markets, in collaboration with economists. Fundamental analysis by econo-physicists can examine the relationship between market efficiency and stability.

3b.7 Career guidance for STEM PhDs

Roughly one-half of PhD holders find their first jobs in non-academic sectors such as non-profits and governmental agencies, corporations, and start-ups.

Recent completers of a PhD feel they received far less information about careers than they needed before entering graduate school. There is a need for increased access to quality career information for prospective and current PhD students.

Though students come to college to prepare for a good job, they don't always know how to find one when they leave. Employers expect preparedness of today's graduates and wish to attract workers with skills for the future. The fresh STEM doctorates need to have a focus on career readiness, one that connects the lab and the classroom to the industry workplace.

Institutions should make career education the focus of quality-enhancement plans, a student-success metric which is a critical part of re-accreditation. An introduction to career development can be incorporated into the first-year seminar, where students connect with a career coach, write a résumé, and start a portfolio of their work. Students should plan for career exploration with flexibility. It is a skill they will need throughout their working lives. The subject matter that the students learn at college and university might become obsolete, but the skills they learn may be durable and lasting.

The universities can also strike deals with employers. For instance, Wipro Limited, an information-technology and business-process-services company, which runs a 10-week course for students, giving them a basic certification in Salesforce, the popular business software. Students get training while Wipro builds connections with students who might be potential hires.

College graduates can no longer rely on their degrees to accurately signal their fitness for a job.

Colleges are therefore, pursuing innovative strategies, including integrating career education into the curriculum, connecting students with career networks, and ensuring access to internships.

A successful research career demands that scientists engage in non-stop learning, as technologies advance, interests shift and discoveries transform understanding, especially in today's era of 'big data'. Training sessions such as workshops, courses or online tutorials can fill the gaps in a scientist's knowledge and skills, helping them to improve and expand their research program and support their efforts to land a job.

Summer placements in a lab can expose students to various new techniques, learning about which is a great way to build skills (and confidence) quickly. Beyond the practical experience, summer placements also introduce students to the reality of working culture both in academia and industry. Often, they are surprised to see that unlike undergraduate study, research is a hugely social and collaborative endeavor.

References

[1] Berkowicz J and Myers A 2014 Throwback thursday: plagiarism vs. integrity *Education Week* https://edweek.org/technology/opinion-throwback-thursday-plagiarism-vs-integrity/2014/07
[2] McPeek M A *et al* 2009 The golden rule of reviewing *Am. Nat.* **173** E155–8
[3] Clancy M *et al* 2023 To speed scientific progress, understand how science policy works *Nature* **620** 724–6

IOP Publishing

Employability for PhD Students in STEM

Jatinder Vir Yakhmi

Chapter 4

Acquiring multidisciplinary skills to beat the 'single-subject cocoon' trap and ability to stay ahead of exponential technologies

By exponential technologies we imply artificial intelligence, ChatGPT, LLMs, IoT, sensors, 5G, 3D printing, machine learning, virtual reality, augmented reality, big data, neural networks, and autonomous electric cars. We discuss how a STEM PhD can learn to convert the above two capabilities to become indispensable for selected jobs/professions.

4.1 Introduction

Scientific research aims to make new discoveries which often lead to: patents, innovations, prototypes, useful commercial products, or new technologies for health, medicine, environment, and energy sectors. That provides a path to new entrepreneurships and jobs.

Entrepreneurship needs a multidisciplinary eco-system, and a multidisciplinary eco-system breeds new innovation, efficiently. What is an innovation? To be called an innovation, an idea must be replicable at an economical cost and must satisfy a specific need, an example being an App-based taxi service like Uber.

4.1.1 Multidisciplinary eco-system for jobs

Most degree holders are graduates in just one subject (physics, biology, computer science, etc). They remain in what may be called a 'single-subject cocoon', which creates a skills gap, which would not be the case if these job aspirants had qualified themselves in at least two non-overlapping subjects.

Along with multidisciplinary capabilities, one needs technical and soft skills to land good jobs, which are being influenced by emerging 'exponential technologies', such as, *artificial intelligence, IoT, sensors, 5G, 3D printing, machine learning, virtual reality, augmented reality, neural networks, autonomous electric cars, ChatGPT.*

A knowledge of STEM is vital in a world increasingly reliant on disruptive digital technologies. Within STEM education we generally count the subjects of science, biology, chemistry, physics, chemical engineering, psychology, computer science and engineering, information science and technology, graphic design, civil engineering, economics, mechanical engineering, electronics and software engineering, and mathematics.

The employment sectors which require STEM skills substantially are: technology, space, medicine, arts, environment, humanity, food, animals, and sports.

New jobs (niche jobs) are more likely to emerge at interfaces (binary or tertiary overlap of the above-mentioned employment sectors). Examples of binary overlap are: (animals + arts), or (medicine + sports), and so on. All such jobs will need talents in multiple subject areas! For instance, a 'food app developer' (binary overlap in food + medicine) would need to have qualifications in biology, chemistry and computer science. Similarly, an 'environment epidemiologist' (in the binary overlap of employment sectors environment + medicine) would need qualifications in biology and computer science and engineering.

Living organisms survive due to homeostatic abilities *(control of body T, BP, pH and sugar levels,* achieved via inter-conversions of chemical and mechanical energy, and self-regulating feedback loops). Our heartbeat, brain waves, pulsed secretion of hormones, cell cycles, biorhythms, are all *autonomous, reversible, with periodical variations, self-regulated beyond thermodynamic equilibrium.*

Homeostasis is one area of science which has potential for jobs for STEM PhDs, and for throwing up new innovations which would bring new technologies. Other areas of science with similar potentials are: materiomics, living matter, wheel-free motion, flying like birds, and microbiome.

To exploit emerging new technologies based on the concepts of nano-, bio-, info-, and cogno- (NBIC), and to generate job opportunities in them needs the development of a synergy between STEM experts with complementary expertise. One example of that is the development of a viable neuro–bionic interface to help cure a brain injury, drug addiction, Alzheimer's, Parkinson's, dementia, epilepsy (the CNS-related diseases, mostly among aged people).

Before moving further, it is inspiring to describe how an expert who has mastered the interface of several disciplines can not only strike it big for himself, but also generate jobs in high-tech areas, for others, too. The example we discuss here is of Sir Anish Kapoor, the most decorated artist alive! He has been very original in applying basic principles of nanoscience, architectures, mechanical engineering, etc, over the last 25 years or so, to produce logic-defying objects which can be huge or intricate, and are for public display. Examples of some of his creations are:

1. In 1992, Kapoor painted walls of a large hole in a floor surface with Vantablack, at Serralves Museum in Porto, Portugal. Vantablack is a black paint which absorbs 99.96% of light and was developed by NanoSystems, UK. Kapoor holds exclusive artistic usage rights to it. (*Vanta: Vertically Aligned Nano Tube Arrays.*)

The exhibit he made appeared to be just a surface painted black, and an unsuspecting onlooker even fell into it.
2. Arcelor–Mittal 'Orbit' created by Anish Kapoor for 2012 Olympic Games in London is UK's tallest sculpture (115 m). It is asymmetrical and has knots and elbows, and onlookers often wonder how it is standing erect, despite these oddities.
3. At a museum in Israel, Anish Kapoor exhibited *'Turning the World Upside Down'*, in 2010, by creating a geometric structure having a highly reflective surface, optically. Passers-by could see themselves in this mirror-like thing, as if walking upside down!
4. Kapoor created LEVIATHAN at the 2011 Monumenta exhibition in Paris's Grand Palais. Creating it meant perfection in computer work, PVC cutting, pre-assembly of the huge Leviathan at a different place and setting it up by a trained crew in Paris. All these steps were done by experts from different European countries and coordinated by Anish Kapoor.
5. He created art beyond physics, called ASCENSION at Rio de Janeiro/Beijing/Venice. It was a safe smoke vortex rising 36 m high, constructed indoors, using the principles of fluid dynamics. The vortex rose all the way to the ceiling. In a way, he 'tamed a tornado' using wind tunnel testing.

4.1.2 AI, ML, and ChatGPT

The cutting-edge 'smart' technologies, artificial intelligence (AI) and the Internet of Things (IoT), have been in use now for some years. For instance, to transform transport management, and ensure that goods reach their destinations at the lowest cost and effort.

In the banking industry, AI tools hold potential to perform in seconds tasks which have been performed until now by a large number of analysts, taking days. The question asked was whether AI will replace those analysts, saving the banks their salaries. The most appropriate reply to this question is that to save their jobs, those human analysts should undergo educative programs in AI, the so-called 'AI literacy' programs already being followed at schools, universities, tech companies and non-profits [1].

AI is essentially a software that can carry out tasks which traditionally required human intellect, such as understanding text, identifying complex patterns, modeling processes, and making predictions. But in an industrial context, AI systems must be engineered for reliability and security, systems which can optimize and improve processes in a variety of sectors like health-care, mobility, power generation, and infrastructure.

ChatGPT, the artificial intelligence language model introduced by OpenAI in November 2022, can respond to complex questions, write poetry, generate code, and translate languages. The widely used chatbot, ChatGPT, was designed to generate digital text, but it was soon discovered that it could do a lot more. Within weeks, it was taught to play Minecraft, a popular video game. The world's leading AI researchers are transforming chatbots into AI agents, which are autonomous system, and could automate almost any white-collar job, and eventually act as personal assistants able to handle a wide range of tasks across the internet.

A wave of excitement created by generative AI that began with the release of OpenAI's ChatGPT continues unabated. Big Tech companies may be secretive about their latest activities, but open-source activity goes on simultaneously, such as that related to large language models (LLMs), the data-hungry artificial neural networks that power a range of text-oriented software, including chatbots and automated translators.

Unlike traditional AI which analyses existing data, generative AI thrives on its training data to produce entirely new and creative text, images, or even audio content. It is already being adopted by various business sectors, especially e-commerce and entertainment. At the heart of generative AI lies 'deep learning', a subfield of machine learning inspired by the structure and function of the human brain. Deep learning algorithms are essentially artificial neural networks with multiple layers that hierarchically process information. This allows the model to recognize patterns and relationships within the complex data.

Generative AI is set to revolutionize the future of work by transforming the way businesses operate and innovate. By leveraging generative AI to drive innovation and efficiency, businesses can unlock new opportunities for high profitability in the digital age.

Generative AI is a technological revolution, but it is not the first one, we have previously had mechanization, industrialization, and digitization. None of them took over all the human jobs, and even if they did consume jobs of a mechanical nature, they also generated new jobs related to automation, etc.

Currently, the advances made by AI are riding on the computer chips called graphics processing units (GPUs), vis-à-vis the workhorse of the conventional computer system called the central processing unit (CPU). The progress made in making faster CPUs suffered a slowdown in the recent past due to Moore's Law, limiting the size of its transistors. GPUs, by contrast, have hundreds, or even thousands of smaller cores, each supported by fewer ancillary systems, such as caches. That allows GPUs to do many simple, repetitive calculations in parallel, much faster than can be done by a CPU.

To power AI products, start-ups and investors, all need GPUs, the critical computer chips. Use of GPUs allows calculations to be run in the fastest and most efficient ways enabling AI companies to analyze enormous amounts of data.

AI learns by absorbing many examples. AI is trained on vast troves of data—virtually all the text, images, and software code on the internet. When prompted, powerful AI chatbots like OpenAI's ChatGPT or Google's Gemini can generate reports and computer programs or answer questions, quickly.

OpenAI has ChatGPT. Google has the Bard chatbot. Microsoft has its Copilots. Not to be left behind, Amazon has built a chatbot, too—an AI assistant named Amazon Q. It has been developed by Amazon's cloud computing division, and is meant for workplaces, and not for consumers [2]. As the leading provider of cloud computing, Amazon already has business customers storing vast amounts of information on its cloud servers. Companies were interested in using chatbots in their workplaces, after making sure that the assistants would safeguard their corporate data and keep their information private. To satisfy that, Q is designed to be more secure and private than a consumer chatbot.

The GPT-4 technology is what researchers call a large language model (LLM). It is an AI system that learns skills by analyzing huge amounts of data.

Adaptive AI in customer service refers to the integration of intelligent systems that continuously learn from customer interactions, preferences, and feedback to tailor responses and recommendations. Adaptive AI represents a paradigm shift in customer service, offering businesses unprecedented opportunities to enhance engagement and satisfaction levels. By delivering personalized experiences, providing real-time assistance, and continuously learning from customer interactions, adaptive AI enables businesses to build stronger relationships with their customers and drive long-term success.

But industry's adoption of AI technology faces challenges, including concerns over accuracy and hallucinations, in which a system provides an answer that is incorrect or nonsensical.

Ever since Microsoft-backed OpenAI's ChatGPT took the world by storm with its human-like responses late last year, AI has moved to the centre of the tech stage. Proponents talk about a quantum leap in industrial efficiency while skeptics have painted a grim picture of millions of people being removed from their jobs.

AI chatbots are powered by LLMs—large neural networks trained on massive datasets of text obtained from the internet, or the Cloud. LLMs do not store data and no longer have access to the training data. Instead, when we build (or 'train') LLMs, they encode content in statistical structures, implying that the text is converted into numbers. Each time we ask a question, the chatbot generates its response afresh.

But, AI chatbots cannot be relied upon to give an unequivocal response since they can produce only probabilistic, and not definitive, answers. Facts cannot be generated from probabilistic patterns.

ChatGPT is a natural language processing tool driven by algorithms, a LLM, which allows people to have human-like conversations with the chatbot. It can answer questions and assist with a variety of tasks, such as composing emails and essays, and writing computer code, among others. By training an AI model it should be possible to identify and locate water leaks, even small ones, not forgetting that about 30% of the drinking water the world produces is wasted due to leaks caused by ageing pipes and ground movements.

AI is being developed to modify the auditory perception of anyone wearing headphones. With AI-based devices you can hear a single speaker clearly even if you are in a noisy environment with lots of other people talking. The system is a sort of real-time training algorithm. The headphones have an on-board mini-computer that runs ML software. The wearer turns towards the sound source and the headphones pick up that source and after only a few seconds, the 'target speech hearing' mode comes in and plays just the targeted speaker's voice even as the listener moves around.

In May 2024, OpenAI launched GPT-4o (symbol 'o' stands for 'omni'), a new version of the AI system powering the ChatGPT chatbot. It has text, image, and audio capabilities designed to increase user engagement and facilitate the creation of new apps. Google too launched its Project Astra AI assistant, just a day later, which has capabilities similar to that of GPT-4o, including a visual memory.

GPT-4o will let ChatGPT talk to users in a lifelike way—detecting emotions in their voices, analyzing their facial expressions and changing its own tone accordingly.

A wealth of open-source technology exists today, including most AI tools—broad ones, such as GPT-4 from OpenAI, and full-fledged libraries for training a specific type of ML model. Big tech makes many of its libraries or task-specific tools available.

One can use AI to help scientists brainstorm, a task that LLMs—AI systems trained on large amounts of text to produce new text—are well suited for. Language models can produce inaccurate information and present it as real, but this 'hallucination' isn't necessarily bad. It signifies a kind of thing that looks real and true.

Machine learning (ML) is a subset of AI that uses data and algorithms to imitate how humans learn. ML is a computational tool that seeks to improve performance for a task using data. The many varieties of ML techniques and algorithms are becoming widespread, impacting multiple social, environmental, economic, and scientific domains.

Many contemporary ML models are so complex that it has become impossible to track the relationship between the inputs and outputs. To tackle these challenges, the field of explainable artificial intelligence (XAI) has emerged which provides robust and reproducible techniques to understand how a model operates.

The number of research papers in materials sciences making use of ML has recently increased phenomenally, including those which deal with applications in materials discovery, property predictions, process optimization, etc.

ML has quietly powered advertising for years, including targeting specific audiences, selling and buying ad space, offering user support, creating logos, and streamlining its operations.

However, it should also be noted that the surge of AI usage has brought to the fore a host of legal and logistical challenges, including the need to protect reputations and avoid misleading consumers.

Google's ML model called GraphCast takes less than a minute to predict future weather worldwide precisely. GraphCast outperforms conventional and AI-based approaches at most global-weather-forecasting tasks. Researchers first trained the model using estimates of past global weather made from 1979 to 2017 by physical models. This allowed GraphCast to learn the links between weather variables such as air pressure, wind, temperature, and humidity. The trained model uses the 'current' state of global weather and weather estimates from 6 h earlier to predict the weather 6 h ahead.

In May 2024, the U.S. National Science Foundation and the Department of Energy awarded grants to several research teams for access to advanced computing resources through the National Artificial Intelligence Research Resource (NAIRR) pilot. The awardee scholars are working in clinical medicine, agriculture, biochemistry, computer science, informatics, and other interdisciplinary fields.

4.1.3 Robots

Home-delivery robots, each able to carry up to 10 kg, serve stores. They navigate along pre-mapped routes using satellite positioning. Sensors, including several cameras and radar, create a shroud of awareness around the robot. On arrival, customers use their phone to unlock the robot's storage compartment and collect their shopping.

A robot, which is driven by algorithms, can answer a maths problem instantly. But it takes a lot of computation power to actually do sensory perception in the environment, such as eye–hand coordination and movement.

Surgeons have been using laparoscopic surgical robots—the kind that don't move on their own but translate the surgeon's movements. However, muscles contract, stomachs gurgle, brains jiggle, and lungs expand and contract—even before a surgeon gets in there. And while a human surgeon can obviously see and feel what they're doing, how could a robot know if its scalpel is in the right place or if tissues have shifted?

Unlike AI or generative art which lives online, robots belong to the physical world, bringing their mistakes and imperfections closer to us as humans. The more information a robot takes in through its sensors, the more efficient and useful it becomes; yet on the other hand, it also raises more privacy concerns.

The approach now gathering steam is to control a robot using the same type of AI foundation models that power image generators and chatbots such as ChatGPT. They can then use brain-inspired neural networks to learn from huge swathes of generic data. They build associations between elements of their training data and, when asked for an output, tap these connections to generate appropriate words or images, often with uncannily good results. A robot foundation model is trained on text and images from the internet, providing it with information about the nature of various objects and their contexts. A trained robot foundation model can then observe a scenario and use its learnt associations to predict what action will lead to the best outcome.

The future of robots and mechanical technology depends on the code that controls them, while their joining with AI, sensor information, independent direction, and human–robot combined effort empowers them to perform more mind-boggling tasks and cooperate with the environment.

Recent robotics are based on AI plus ML. This is because these innovations empower robots to gain information so they can recognize examples or even settle on choices without human intercession.

A robot does exactly what it's told to do, eliminating the chance of human errors, the proof of which is that the number of airplane crashes have reduced sharply since the autopilot was introduced. Similarly, a robotic arm with a gripper can and does approach the versatility of human fingers.

The use of robots by big manufacturing companies, retailers, and movers of goods rose sharply after 2019, during the pandemic, when human workers became scarce due to lockdown.

We could theoretically hand over complete decision-making to the autonomous robots, such as in a driverless car, but what happens if something goes wrong, like the car has an accident? A human should be around in a supervisory role, reviewing and standing by in case of emergency.

We should ultimately have an inherently safe system, such as a soft robot, steerable by externally controlled magnets, designed to snake deep into say a patient's lungs to view the tissue there, or through the intestine to look for early signs of colon cancer—a robot that can navigate the narrow passages on its own, eliminating the need for x-rays to help guide a human operator.

By combining data with the huge amounts of text used to train chatbots like ChatGPT, one can build AI technology that gives its robots a better understanding of the world around it. After identifying patterns in images, sensory data and text, the technology gives a robot the power to handle unexpected situations in the physical world. That points to a role to be played by STEMM PhDs, who as AI researchers branching into robotics, can eventually create a robot that is more autonomous and adaptable across a wider range of circumstances. Imagine starting with the available technology of robot arms that can 'pick and place' any factory product, evolving into humanoid robots that provide company and support say for older people.

Human intelligence blended with AI will bring a paradigm shift in imparting education. Humanoid robots will assist the teacher in teaching, while the teacher can build competencies for innovation. Teaching robots are capable of teaching students in schools, in the classroom along with the human teacher and also in a stand-alone mode. For instance, robot teachers work along with teachers in delivering lessons at classes 7, 8, and 9 for physics, chemistry, biology, geography, and history. They clarify doubts by answering questions and conduct an automated assessment at the end of the class with the help of analytics.

4.1.4 5G technology

5G mobile networks have the capability to transfer data at more than a gigabit per second, versus 4G networks that typically offer speeds of around 50 megabits per second. Engineers hope that 5G will eventually connect more than just phones. It could also link sensors embedded in everything from farm machinery to medical devices, forming the so-called Internet of Things which has been difficult with 4G. Phones that support 4G will continue to work for a while, because 5G isn't replacing 4G, it is being built on top of it.

The true purpose of 'the cloud' is to float unseen all around us, silently, creating an ever-present connectivity. The cloud is a system of millions of hard drives, computer servers, signal routers and fiber-optic cables. These elements act like the water droplets, ice crystals, and aerosols that make up a true cloud. They are nebulous. They are constantly shifting. But they are in close connection with one another over large distances of space and time.

4.1.5 Resource-guzzler AI

Just a decade ago, data centers drew 10 megawatts of power, but 100 megawatts is common today. Right now, AI is only a small percentage of the global data centre footprint, but it will skyrocket from the 2% of today to 10% of global power use by 2025.

AI technology can be used to moderate cooling loads and adjust energy use in concert with changing weather, which could make a data centre more efficient, perhaps by 10%. AI is already helping manage the environmental impact of another key part of modern society's infrastructure: data centers. Their energy usage is significant, much of it used to keep their thousands of servers cool.

The vast new data centers of big tech are run at huge cost to the environment. The infrastructure used by the cloud accounts for more global greenhouse emissions than commercial flights. Large language models such as ChatGPT are some of the most energy-guzzling technologies of all. It seems that about 700 000 l of water could have been used to cool the machines that trained ChatGPT-3 at Microsoft's data facilities.

Newer AI models are getting bigger. Bigger models require the use of more and more powerful GPUs, and take longer to train—using up more resources and energy, and also water in the creation and use of various AI models. Data centres use water in evaporative cooling systems to keep equipment from overheating, during the training of say GPT3.

Furthermore, while minerals such as Li and Co are most commonly associated with batteries in the automobile sector, they are also used in batteries used in data centres. The extraction of these often involves significant water usage and can lead to pollution, undermining water security.

4.2 Influence of AI on jobs

An oft-repeated question is: will artificial intelligence put a large number of people out of jobs? Let us recognize that the same doubt was raised earlier on the arrival of robots and automation which eliminated only blue-collar and other physical work. ChatGPT, as an AI innovation, is out to target jobs that encompass non-routine, cognitive tasks [3]. As per some predictions, 300 million jobs are at risk, worldwide, which include some already gone as 'lay-offs' during the pandemic, etc. Examples of jobs that are presumed to be at stake are tech jobs such as computer programmers, coders, software engineers, data analysts, and media jobs, such as graphic designers, advertising, content creation, technical writing, and journalism.

But the general answer to the question above is NO. If technology permanently puts people out of work, then why, after the introduction of so many of new technologies in the past, are there so many jobs left? New technologies allow people to enter new fields—like the shift from agriculture to manufacturing. High technology like AI requires higher education and expertise to unlock its productivity benefits, so obviously STEM PhDs should be more in demand, if they can adapt and learn new knowledge and skills required for the job opportunities.

Consider a researcher whose job includes the collection of data, performing data analysis, and writing reports. With the advent of AI, the technology might take over the data collection task, leaving the researcher to do data analysis and write the report. The researcher might, in fact, employ AI to do their data analysis, and thus have value added to their job. It would, in fact, be beneficial from the perspective of a STEM PhD to have machines take over low-value work—as long as the ever-increasing specialized expertise of the researcher enriched by learning of new skills allows them to perform the higher-value tasks.

Proficiency in coding skills isn't enough in today's job market. With the rise of AI, coding has been rendered as a complementary skill to AI expertise. AI specialists, who can code and understand AI algorithms, are in high demand. AI-driven changes in employment have become a necessity in the face of a rapidly evolving job market.

As time goes by the list of potential tasks to be handled won't remain static, and its ever-evolving nature will pose new problems to solve, which a STEM PhD with analytical capabilities would be able to handle, hopefully more expeditiously than others. AI does have the potential to improve our lives, vastly, in several sectors. The key is to make it work for us.

Thanks to the computers and the internet, a massive amount of data is continuously generated in today's world. In recent years, there has been a lot of hype around 'big data'. Data science is now emerging as an independent discipline, extending the field of statistics to incorporate advances in computing with data. In essence, data science is a mix of statistics, mathematics, algorithms, engineering prowess, and communication and management skills. People want to analyse 'big data' to create effective strategies in different aspects of human life—in business, industry, sports, health-care, or policy-making. Therefore, the new role of Data Scientist has emerged which refers to professionals armed with expertise in the topics mentioned, and with the training and curiosity to tackle the world of big data. The key skills of data science involve a mix of coding, statistics and machine learning, and business acumen and communication.

4.2.1 AI engineers

AI has applications across tech innovations like the IoT, AR/VR, robotics, ML, etc. AI-related careers are considered future-proof because they help AI in transforming the global economy. Enrolling in a certified AI engineer course will enhance expertise, aptitude, and talent in the emerging AI and ML fields.

The AI engineer certification expects you to learn AI on the cloud, machine learning algorithms, Python and NLP fundamentals. In addition, as an AI engineer, you must acquire a blend of soft and hard skills. Hard skills relate to mathematics, computers, and engineering. Coding, algorithms, statistics, and big data technologies are especially crucial for AI engineers. Expertise is also required in software languages, cognitive language theory, and mechanics, linear algebra, probability, statistics, robotics, machine learning, data structures, and logic.

The soft skills required for AI engineers relate to solving problems and communicate, viz. verbal and written communication, teamwork, active listening, critical thinking, creativity, and time management

An AI engineer is a data engineer who excels in creating AI-powered applications, focusing on maximizing model capabilities and optimizing workflows for LLMs.

Further, we could have the post of an AI researcher, which could be an AI/ML researcher, a data scientist, or a research scientist.

Finally, there is the role of prompt engineer who designs and refines the inputs (prompts) given to AI models, like ChatGPT. Prompt engineering is particularly relevant when using ChatGPT or similar models which use a conversational approach. It involves understanding the capabilities and limitations of the AI model and applying strategies to guide the AI towards providing the most accurate, relevant, or creative responses. The new job of AI prompt engineer is a high-salary job that helps guide generative AI [4].

Can humans be dispensed with totally by an AI chatbot, which writes proposals for the clients of a company? For that, the chatbot is fully geared to scan through thousands of files and fish out relevant information to generate proposals, and in the process, save the company time. The answer to the question is that the company still needs to check the chatbot output. This is because chatbots do not have human intelligence. They can have great efficiency on augmenting productivity through assisting in writing tasks, creating original text, and helping a human to explore new issues, through discussions, but can't replace human expertise.

4.2.2 Will AI replace humans or complement humans?

There is a certain awe, a deep fear among the regular users of the internet that AI, particularly in its ChatGPT mode, unveiled recently, may take away human jobs significantly.

Humans have always been stunned by new technologies which suddenly quickened the pace of numerical calculations, or speed in communication of data. For instance, the slide rule, or the Friden machine calculator, both of which were non-digital tools that were new in 1960s, did make an impact by providing much faster calculations than ever before. Then came in 1980s the hand-held digital calculators, at least for the four basic arithmetic functions, viz. addition, subtraction, multiplication, and division, although several other functions including algebraic, logarithmic, and trigonometric were also incorporated into the hand-held calculator. More recently, instant communication by fax or email received public acceptance, just before the arrival of the smartphone which offered audio and video calls and GPS facilities. Still, humans could not be side-lined by such technologies.

With great expectations from generative AI tools like ChatGPT and Google's Bard, the big tech and venture capitalists in Silicon Valley are investing billions of dollars in the technology.

Some experts have predicted that AI will displace people from their work, while others have said the tools will augment individual productivity [5].

Generative AI is more likely to augment than destroy jobs by automating some tasks rather than taking over a role entirely. Generative AI's potential lies not in replacing humans but in assisting humans in their efforts to create hitherto unimaginable solutions. Knowledge work primarily involves cognitive processing of information to generate value-added outputs. In marketing and advertising, generative AI is being considered as a tool to automate routine content generation such as creating product brochures or personalizing email campaigns, reducing the cognitive load on knowledge workers, and thereby freeing their mental capacity to focus on higher-value unstructured tasks.

AI isn't coming to take your job; it's coming to change your job. The winners will ride that wave, and benefit, as against those who refuse to join who may be the losers. Technological revolutions in the past have created more jobs than they have destroyed. The advent of the calculator didn't make math less important, but it changed how we taught math in schools. The calculator became a useful tool for rocket science because of its ability to structure a problem as a set of mathematical equations. Just as math education changed with the arrival of the calculator, our overall approach to education needs to adapt to AI technology.

AI does not have emotional intelligence, humans do. Hence, humans can build meaningful connections and empathize with others, a crucial factor for business growth. Also, AI is limited by the data it receives and needs to adapt to new situations. Human reasoning and creativity cannot easily be replicated by machines. AI can't perform manual labor. AI has no soft skills, such as teamwork and effective communication, which gives humans an advantage over AI. Thus, AI is meant to complement human ability and intelligence, not compete with it. Humans must learn to work with AI, not fear it. A prudent approach for knowledge workers is to harness the potential of AI as a complementary tool, amplifying their capabilities and adapting to the evolving landscape of work.

AI algorithms imitate real-world systems. The more repetitive a system is, the easier it is for AI to replace it. That's why jobs in customer service, retail and clerical roles are often named as being the most at risk. Let us take an office clerk as an example, whose responsibilities include answering phones, taking messages, and scheduling appointments. We now have access to AI tools that can perform all these tasks. They can also work non-stop, without the need to take breaks. ChatGPT can replace customer service representatives, data entry clerks, and copywriters. AI could also replace technical writers, translators, and interpreters.

The skill to understand and work with AI will soon become as routine as working with PCs is. Generative AI, such as ChatGPT, will be integrated into several tools employed by knowledge workers, automating routine tasks like note-taking, and drafting personalized customer messages. Automating these tasks will allow knowledge workers to concentrate on value-added activities where human expertise is indispensable, such as looking for nuance in context, exercising emotional intelligence, addressing moral and ethical considerations, and fostering creativity and innovation.

4.2.3 Which human skills AI is unlikely to master?

Three human skills that AI can't replace are curiosity, humility, and emotional intelligence. Every new technology takes time to be useful in the long run to complement human capabilities, but the technologies themselves cannot outsmart humans. For example, emotions, including nuances of social and cultural parameters used conventionally by humans, cannot be exercised by chatbots or AI, or machine learning. Humans historically have a proven capacity to match and then master new technologies. AI can only change the nature of the jobs held by, say a STEM PhD, but not take the jobs away. If a STEM PhD can keep upskilling themselves, then the new emerging disruptive technologies would only be a boon to them as they can employ these technologies like AI/ML to take up bigger challenges.

The primary skills of AI are technical, viz. coding, calculation, and math. What AI cannot do is offer human skills, including being creative, wise, and having critical judgment. An employee with these skills can therefore distinguish themselves. Analytical and creative thinking remain the most important skills for workers. So, if you're worried about keeping your job in the future, it's worth acquiring more of these skills.

Erudition, a verbal skill, is the exclusive and irreplaceable ability of humans using which they can think and speak clearly on the spot, in a manner that can be compelling and persuasive.

AI will provide the intelligence, the computational speed and scale to operations driven by automation. But AI does not invent; it just predicts on the basis of past patterns. It is marketers who invent, and the role of AI is to learn what works, for whom, when, and how.

Even as AI moves to acquire most jobs that involve repetitive actions, the qualities which makes us human will become much more sought after in the employment market. And these qualities include courage, vision, wisdom, and empathy.

Human users will always have additional knowledge or context which is proprietary, that the AI doesn't, and hasn't been trained on that knowledge.

AI models can and are likely to make mistakes, in logic, efficiency, or inference, even if infrequently. Human skills to test, edit, double-check and innovate will always remain valuable.

A human-shaped robot will be able to physically interact with the world in much the same way that a person does. However, controlling a robot is incredibly hard. Simple-looking tasks, such as opening a door, are actually hugely complex, requiring a robot to understand how different door mechanisms work, how much force to apply to a handle and how to maintain balance while doing so. Giving AI systems a body can cause mistakes and even threats to the physical world. A robot making a wrong decision can physically harm you or break things or cause damage. While automation has many benefits, mainly standardization and repetition, a robot can't create the kind of uniqueness that comes from the human touch in various artforms, for instance.

4.2.4 With some efforts STEM PhDs can remain unbeaten and stay ahead of AI

STEM PhDs should look at their unique qualities which make them indispensable, such as, critical thinking, which makes humans understand new problems from various angles. Human qualities such as making friends and wishing to work with them are not mechanical, and so will enable humans stay ahead of AI. The quality of liking to work with friends can make a STEM PhD more productive and push a company to grow faster.

The types of skills needed are constantly evolving. Hence, many employers prefer workers with a strong foundation in a variety of skills. Analytical and technical skills are still important. Employers are looking for also what are called foundational skills. These not only include social skills, but also analytical thinking and math.

Building things in the AI field isn't very complex. If you're already good at math, engineering, and business, there are few limits to what you can do. Software engineers could let AI tools autogenerate programming code for basic purposes, leaving them to write code more efficiently while spending more time on other activities, like system design.

There are no shortages of client demands for digital transformation, emerging AI applications, renewable energy, risk mitigation, pricing/yield management, climate change, and supply-chain realignment, to name but a few. As an applicant, a STEM PhD needs to articulate how their background and experience will help the firms fulfill specific client needs. When applying you should focus on your unique strengths; think about what is it that you're uniquely able to contribute.

4.3 Beating the 'single-subject cocoon' trap for enhanced employment opportunities

UCLA has created a course called Mathematics for Life Scientists that covers classic calculus topics such as the derivative and the integral, but dealing with their application in a biological context. Biology is rich in patterns, and mathematics is the science of patterns. An emerging field called mathematical biology has potential to revolutionize microbiology, biotechnology, evolutionary biology, and healthcare. Mathematical biology will help build mathematical models to describe a biological cell, called the 'whole cell model', which will allow us to compute the life of a cell and help us understand how the human body works.

Whole cell models will also let us understand how biofilms form, and learn to inhibit biofilm formation, to bringing back the purity and integrity of the water supply for drinking and for agriculture.

But where are the jobs? A big reason for not securing jobs is the skills gap, arising from candidates (even STEM PhDs) often falling into the 'single-subject cocoon', meaning they have expertise in just one subject, say only biology, or only physics, or only mechanical engineering, etc. One needs skills and multidisciplinary capabilities to land good jobs, which are being influenced by emerging exponential technologies: AI, ChatGPT, IoT, sensors, 5G, 3D printing, machine learning, virtual reality, augmented reality, neural networks, autonomous electric cars.

Well-informed students are now doing a Bachelor's degree in physics and cognitive sciences, or, an MSc in mathematical modeling and scientific computing, or, an MSc in international health and tropical medicine. Bioinformatics, which applies data science to biological data, is another favorite.

We attempt to underline the potential of multidisciplinary research activity by citing two classical examples of scientists who achieved great heights in their chosen professions:

1. Professor Mauro Ferrari (b. 1959) did his Laurea in maths (Padova, Italy), MS and PhD in mechanical engineering (Ucal Berkeley), and rose to become President, CEO, and Director of The Methodist Hospital Research Institute, Department of Nanomedicine, at Houston (USA). Before that he was Professor of Biomedical Engineering, Internal Medicine, Mechanical Engineering, and Materials Science at Ohio State University. His research focuses on understanding the physical and biomechanical barriers that reduce the efficacy of cancer therapeutics. A new drug developed by him, called iNPG-pDox, showed better results at lower doses compared to standard for metastatic breast cancer. His favorite research areas have an impressive multidisciplinary range, viz. nano(micro)technology, physical sciences, math, biomechanics and materials science, all of which he uses to develop technologies for drug delivery and cancer therapeutics.

2. The 2009 Nobel Prize winner in Chemistry, Professor Venkatraman Ramakrishnan undertook a BSc in physics at the University of Baroda (India), and followed this up with a PhD in physics from Ohio University, after which he graduated in biology from the University of California (San Diego). Thereafter, he worked for 12 years on ribosomes at Brookhaven National Lab (USA). Currently, he works in the Molecular Biology Lab at Cambridge (UK). He received the 2009 Chemistry Nobel for *'the protein producing ribosomes, which transform DNA into LIVING MATTER—like skin and immune system …'*, as stated in the citation.

Living matter (sometimes also called active matter, or self-propelled matter), lies at the multidisciplinary interface—a confluence of physics, chemistry, biology, and materials science/engineering. In other words, it deals with science, technology, engineering, and mathematics (STEM).

4.3.1 Role of multidisciplinary expertise in job opportunities, typically for STEM PhDs

ML has huge capabilities as a method for data analysis, but it is extremely 'data hungry', with unstable performance unless the size of the training set is much larger (20 times or more larger) than the number of attributes, a condition that is seldom met in biomedical engineering studies. In many research papers published in this area, it is difficult to know for certain that the test data were rigorously separated from the training data, or that the analysis was not adjusted during the study to obtain the best results, both being necessary prerequisites for a valid study. Perhaps very few people have tried to replicate many of the ML studies, because replication studies are contentious, expensive, and difficult without clear resolution. Replication is important, but generalizability of knowledge is even more important.

The AI boom is based on neural networks and machine-learning algorithms, which first grew gradually since 2010 and then very sharply. And this boom includes AI coding projects for sharing code which has seen an exponential growth, with most machine-learning being done in industry. STEM PhDs have to take note that their kind of academic expertise is being more and more deployed to analyze the models coming out of companies, to look essentially for weaknesses in them.

An excellent example of multidisciplinary strengths is the Institute for Bioengineering of Catalonia (IBEC), Barcelona, which focuses on controlling biological processes. It has an interdisciplinary Lab to host joint activities of researchers from physics, chemistry, biology, medicine, and engineering who are engaged in microfabrication, microscopy characterization, biomodelling, and 3D bioprinting units. A clinical mentoring program connects PhD students with a clinician [6].

Making and designing new chips, which provide the heavy processing power needed to develop new applications, is a multidisciplinary activity. It is estimated that developing ChatGPT took 10 000 chips supplied by the US chipmaker, Nvidia.

The AI industry is likely to consume 3.5% of global electricity by 2030. That itself is reason enough to usher in a transition to green tech, to keep its carbon footprint under control.

4.3.2 Electric vehicles and batteries—opportunity for jobs for STEM PhDs

The fourth Industrial Revolution is well underway. Automobile manufacturing will require fewer parts and assemblies and there will be much lower service/maintenance requirements.

Plug-in hybrid models of automobiles, which can travel on just electricity for more than 40 miles and have a gasoline engine for longer trips, have much smaller batteries than electric vehicles and can be recharged relatively quickly. But these vehicles, may not be as financially or environmentally viable when driven long distances on just gasoline.

A transition from fossil-fuel based internal combustion engines to climate-friendly electrically driven vehicles will provide the much-needed push towards climate-friendly green technology, and create new jobs by producing electric cars. That should encourage young people to study subjects like batteries, robotics, and coding. Electric vehicles (EVs) need axles but typically do not need long drive shafts because the motors can be placed close to the wheels. EVs use a different technology, and therefore, have fewer parts inasmuch that EVs have no cooling system, no exhaust, no transmission. Brake life is increased as regenerative braking slows the vehicle and charges the battery. Even if the transition takes the hybrid route, EVs are likely to generate job opportunities for several years.

Hydrogen (H_2) fueled cars might eventually take over from EVs because H_2 cars can be filled in minutes at current gas stations refurbished for handling H_2 fuel, instead of requiring hours to re-charge batteries. Hydrogen is the fuel used to create electricity to power the electric motor. That way, H_2 powered vehicles are also EVs, if one considers that energy (fuel) stored in a hydrogen tank or energy stored in a battery, both provide power to the same electric motor. Battery manufacturers are switching over to lithium-iron phosphate (LFP), which requires no cobalt, and already scientists are developing even cheaper Na–Fe–P batteries (sodium replacing lithium). Obviously, there is scope for jobs for STEM PhDs in this sector.

Electric vehicles are so far not being sold in as large quantities as expected because they are more expensive than combustion or hybrid models, the lack of charging stations for electric vehicles, their range and performance in cold weather have also caused some people to hesitate.

Besides, internal combustion engine (ICE) vehicles, too, have been getting better and better over the years. Defects per vehicle and money spent on maintenance and repair has been declining, so much so that it's routine that an ICE vehicle is expected to hit 100 K miles before any major maintenance is required. EVs do better since there far fewer things to maintain.

EVs are high-quality, high-performance, zero-emission, and quiet in their operation, with driving ranges of more than 300 miles, recharging times of less than 30 min and batteries that can be recycled. An EV is also described as a computer with wheels. A further new innovation to come for EVs is a heat pump that greatly improve distance for vehicles in cold weather. But, above all, dependable, and widely distributed public charging stations are the key driver to scaling the adoption of EVs, and ushering in zero-emission EVs, significantly reducing overall carbon use. EVs convert the electrochemical energy stored in the battery to motion using motors that have efficiencies of over 85%. Even accounting for the losses of delivering energy to the charger, EVs remain more energy- and carbon-efficient than fossil-fuel vehicles. And electrification means greater use of renewables as the grid continues to decarbonize. That would mean more jobs in the EV and battery sectors.

There is, no doubt, an urgent need to drastically cut carbon emissions, for which the world should ditch the ICE as early as possible. Completing the transition from ICE to EV will be challenging and will require imagination, innovation, and investment. That points to the job prospectus for STEM PhDs, because, we need new battery chemistries to extend vehicle ranges at lower costs; to understand and implement environmental impacts of mining and battery production [7].

Because electric vehicles have no emissions, they reduce urban pollution, which in turn reduces asthma and other respiratory problems that cost many lives and resources. However, we need to make electric cars safe, too. There should be a mandatory audible warning systems for electric vehicles, emitting sounds to alert pedestrians of an approaching vehicle. This is already being done although we need to make the sounds louder. We need to educate both drivers and pedestrians about the specific risks associated with the quieter nature of electric and hybrid vehicles, while drivers should exercise caution in pedestrian-dense areas.

Wirelessly connected to traffic lights and the surrounding streets, the driverless car avoids collisions and reduces congestion. Humans have a tendency to over-rely on automated systems, such as autonomous vehicles (AVs), and this automation bias is a hard habit to break. We tend to perceive technology as infallible viz. our inclination to blindly follow navigation systems, like GPS. If we want EVs to drive as well as an alert and experienced driver does, we might want to set the bar of EV collisions much higher than the national average.

4.3.2.1 Batteries and lithium
Following the rapid expansion of EVs, the market share of lithium-ion batteries (LIBs) has increased exponentially and is expected to continue growing, reaching 4.7 TWh by 2030 as projected by McKinsey [8].

The holy grail of EV technology is a 745-mile capable solid-state battery that can fully charge in 10 min and is twice as energy-dense as current cells! Current cells store their energy in liquid or paste-like electrolytes, whereas solid-state batteries use solid electrolytes, which are far more energy-dense and lightweight, so they can contain more energy per volume. Solid electrolytes also have lower electric resistance than liquid ones, leading to far faster peak charging speeds. Solid-state batteries have the potential to be far cheaper and more eco-friendly than present cells. They also resist battery degradation, meaning they could last far longer than current cells.

To mitigate the mounting threat of climate change as demonstrated by the intense heat waves in 2024, the world needs to abandon fossil fuels and adopt lower-carbon technologies. A key to reducing emissions to cap the global temperature increase as per the Paris Agreement of 2016, is access to rare earth elements (the Lanthanide series in the Periodic Table), and lithium. The demand for all of these is increasing the world over. China accounts for more than half of the global production of rare earths.

There are three routes to obtaining rare earth elements: (i) by extraction after mining directly from the earth, provided the producing country has access to its own mines; (ii) by recovery from secondary sources, such as end-of-life electronics; and (iii) by extraction from industrial wastes like coal ash and waste products from mines.

The chemical process technology that extracts rare earth elements from coal ash leaves behind a solution rich in rare earth elements, plus a residual solid containing iron and other metals. Subsequent processing transfers rare earths into an ionic liquid—a salt in liquid state. After reducing the amount of iron in the solution, through repeated reduction and leaching processes, one obtains a rare earth-rich solution, containing several rare earths which have then to be separated to get individual rare earths as pure metals or oxides.

There is no dearth of coal ash, but it needs to be handled with care. Regulatory action is required to make sure that the by-products are disposed of appropriately. The governments should ensure that communities receive adequate notice of nearby extraction activities, too.

The battery—the anode and the cathode—have chemical reactions that are slowed during extremely cold temperatures, affecting both the charging and the discharging of the battery. Charging stations see longer queues in winter, as vehicles take longer to charge in colder weather.

Lithium metal can store large amounts of energy in a small space, which is why it is attractive as a battery material. But that also means there is much energy available to turn into heat and even flames in case of a short circuit. Lithium battery fires are a grave concern for battery manufacturers. A fire at a lithium battery factory near Seoul in June 2024 killed 22 workers.

4.4 For continued employability STEM PhDs should upskill their job skills regularly

In view of the potential impact that the new generative AI application, ChatGPT-4 (Chat Generative Pre-Trained Transformer-4) may have on the job market, a candidate should focus on upskilling for newer roles, rather than face job security threats.

ChatGPT-4 has the potential to enable existing jobs to be performed more competently and efficiently, making the existing methods often obsolete. Adapting to new skills and embracing new challenges is essential for employees to perform proficiently, using AI tools. In addition to enhancing job efficiency, AI technology can also generate new employment opportunities without compromising the primary objectives of existing jobs.

Important thing to know about upskilling is that every employee needs to be doing it all the time. Certifications will show that you have attained knowledge and capabilities in a certain field. MIT OpenCourseWare's YouTube channel inspires millions of learners across the globe to expand their knowledge and develop new skills for free.

In an age of disruption, the only viable strategy is to adapt because obtaining a new skill gives you, on average, just about 5 years of advantage. So, upskilling for employability has to be a continuing process. The knowledge economy, is now giving way to a relationship economy, where people skills and social abilities will become a greater key to success than ever before. Simultaneously, job aspirants must work on gaining the technical skills which help in getting a good job.

STEM PhDs should go for AI to meet their diverse needs as learners, but ensuring simultaneously that foundational academic skills are complemented with human skills such as adaptability and curiosity, and the ability to learn in teams. It is important to learn and strengthen individual skills such as empathy, understanding, and the ability to build connections across diverse perspectives. Plan to go beyond the traditional metrics of success, and stay competitive by becoming a lifelong learner. Be advised that to gain useful knowledge to stay employable, just after earning a PhD, and later all though your life, you should practise curiosity and humility.

The MicroMasters in Statistics and Data Science was developed by the MIT Institute for Data, Systems, and Society and *MIT*x. As a part of MIT Open Learning, the MicroMasters programs have drawn in almost 1.4 million learners, spanning nearly every country in the world. More than 7500 people have earned their credentials across the MicroMasters programs, including: Statistics and Data

Science; Supply Chain Management; Data, Economics, and Design of Policy; Principles of Manufacturing; and Finance [9].

In 2023, the Government of India launched 'AI for India 2.0', a new program that provided free online training. Several online certificate programs on AI, ML are available. Microsoft has a 12-lesson course on Generative AI for Beginners, Harvard Business School offers an AI course, and upGrad and Coursera have courses on AI where participants can develop job-relevant skills with hands-on projects. Candidates now have diverse career options in AI and ML, including roles as research engineers, business intelligence analysts, data scientists, and ML engineers, with all IITs in India offering dedicated courses in these fields.

Generative AI, has huge skills in writing, programming, and translation. Skills that any job requires, are likely to be affected by generative AI technologies. Nearly all of the skills that a software engineer has at present, say in programming languages can be replicated by AI. Skills associated with jobs in legal and finance sectors also are likely to be highly exposed. Advances made in large language models, as in the GPT, allow for the generation of text that is almost indistinguishable from that which is written by a human. With human-like conversational ability and personalities, AI agents can support humans in roles such as a personal assistants, or mental health counselors.

As AI makes its way gradually to social settings, we need to check if a robot can handle social intelligence. We'll face a host of new ethical questions. For instance, can a robot apologise with a human touch, that too with a socially innovative manner, as per the occasion.

4.5 New jobs to be created by AI

Technological advances in AI are not moving to make us jobless, and will instead create more opportunities for specialized 'niche' jobs. AI can create jobs in high-tech fields like computer science, data science, machine learning, robotics and automation, math and statistics, and business. Experts note that AI will create many employment opportunities, including jobs that don't yet exist. Job losses will be partially offset by job gains, say for machine learning specialists, and emerging new jobs like prompt engineers.

Sector-wise, some new jobs that are likely to emerge are listed below. To take advantage of new openings, STEM PhDs should embrace AI technology tools to enhance and improve their workforce's capabilities rather than fear them as a threat to their job security.

Autonomous vehicles: Autonomous Systems Designer, Autonomous Vehicle Operator

Health sector: Digital Detox Specialist, Personalized Nutritionist, Digital Health Record Technician, Digital Wellness Coach, Medical 3D Printing Specialist

Virtual and augmented reality: Virtual Event Planner, Virtual Interior Designer, Virtual Reality Developer, Augmented Reality Product Specialist, Virtual Personal Shopper

Virtual identity and security: Virtual Identity Protection Specialist, AI Security Analyst, Data Protection Officer, Personal Privacy Consultant, Digital Identity Manager, Blockchain Developer, Cybersecurity Analyst

AI and machine learning: AI Training Data Creator, AI Ethicist, AI Product Manager, AI Content Curator Specialist, AI-Language Translation Specialist

Smart homes and cities: Smart Home Technician, Smart City Planner, Climate Change Analyst, Sustainability Manager

Robotics: Chatbot Personality Designer, Social Media Behavior Analyst, Human–Robot Interaction Specialist, Data Visualization Specialist

4.6 Typical examples of new emerging job opportunities for STEM PhDs

The use of AI in science has been rising in science applications, such as exploration of new materials, and rapid weather forecasting. The most sought after skills for decades, viz. technical and data skills, also happen to be most vulnerable to advances in AI. But, luckily, people skills, long undervalued as 'soft,' will possibly remain the most durable.

In what follows, we list 12 examples of new job categories that are likely to arise in different sectors of human endeavor and industry, which have a direct link with important societal ecosystems viz. energy, climate, health, and employment. These new job categories should prove valuable for STEM PhD candidates, as most of them draw on capabilities of AI, multidisciplinary expertise, technical skills, and soft skills.

4.6.1 AI and ML in agriculture

ML/AI technologies have the potential to help implement precision farming by creating algorithms uniquely suited to specific farms and locations. New jobs are likely to emerge in agriculture sector in the areas of (i) ML within crops, pasture, and irrigation; (ii) ML in predicting crop disease; and (iii) society and policy of ML [10].

4.6.2 AI in radiology

In the field of radiology, we benefit by the application of intelligent algorithms and ML techniques to analyze medical images such as x-rays, CT scans, and MRIs. These algorithms are trained on massive datasets of labeled medical images, allowing them to identify patterns and anomalies that might be missed by the human eye. Once trained, the AI can analyze new, unseen medical images and carry out image segmentation and image interpretation. It can automatically detect abnormalities, highlight areas of interest, and even suggest potential diagnoses based on its learnings from the training data.

4.6.3 Using machine learning to identify new deposits of in-demand minerals

By employing ML to characterize patterns embedded in the multidimensionality of mineral occurrence and associations. Mineral association analysis quantifies high-dimensional multicorrelations in mineral localities across the globe, enabling the identification of previously unknown mineral occurrences, as well as mineral assemblages. Using AI, therefore, one can predict new deposits of critical minerals, specifically rare earth element (REE)- and Li-bearing phases. Mineral association analysis has scope as a predictive method to enhance our understanding of mineralization and mineralizing environments on earth, across our solar system, and through deep time [11].

4.6.4 Thin films for computer chips

Recent development of thinner films, just a few atoms thick, of transition-metal dichalcogenides (TMDs), hold promise of building more compact and powerful computer chips.

4.6.5 Designing medicines that can go across the blood–brain barrier

The blood–brain barrier (BBB) is designed by Nature to protect the brain from toxins in our food, and harmful infectious viruses from entering the brain. But the same BBB also stops many medicines getting into the brain, and thus poses a challenge and an opportunity for development of new medicines, which can be designed to go across the BBB.

Recently, scientists at the University of Texas at Austin have successfully delivered fluorescent sensors across the BBB to monitor neurotransmitter levels in the brain, which could specifically advance the diagnosis and treatment of Alzheimer's disease. The method involved encapsulates ATP (adenosine triphosphate) aptamer sensors in exosomes, which are able to cross the BBB and provide real-time images of neurotransmitter levels. Sensors accumulate in the brain and identify low levels of ATP in areas affected by Alzheimer's [12].

4.6.6 Neuromusculoskeletal prosthesis

A highly integrated bionic hand with neural control has been developed which is connected directly to the user's nervous and skeletal systems [13]. This will give a big impetus to upper-limb prosthetic devices aiming to restore function which vary in their degree of anthropomorphism, from hooks and grippers to hand-like robotic devices, and will definitely open the doors of job opportunities for STEM PhDs with multidisciplinary aptitude.

4.6.7 Programmable bacteria to kill cancerous tissue

Investigators are developing synthetic programmable bacteria to help kill cancerous tissue [14]. By harnessing the power of beneficial microbes to modulate the immune system, biomedical engineering has the potential to alter the future of medicine for autoimmune diseases such as type 1 diabetes and rheumatoid arthritis, and in the process create new multidisciplinary jobs for STEM PhDs.

4.6.8 Quantum technology

PhDs holders appear to have a monopoly right now on jobs in quantum technology. Quantum technology includes magnetic sensors and atomic clocks, as well as quantum computers (QCs). Touted as a paradigm shift in technology, these are devices in which quantum mechanics enables extremely precise measurements and a fresh way for computers to crunch numbers. The fast-emerging field of QCs which will unfold hugely enhanced computing capacities in a not too-distant future will use quantum entanglement.

Unlike regular desk-computers, which encode words or numbers as collections of 1 s and 0 s called 'bits,' QCs rely on quantum bits or 'qubits,' which assign weights to their 1 s and 0 s, which means there is a probability associated with measuring either number. They embody a bit of both states until you measure them. Quantum algorithms run on these qubits, and, theoretically, perform calculations, causing their probabilities to interfere and increasing their odds of finding the ideal solution. That makes QCs very fast, and capable of cracking hard problems encountered in the subject areas of drug discovery, AI, etc. Pharmaceutical companies and EV manufacturers have begun to explore the use of QCs in chemistry simulations for drug discovery or battery development. To elaborate, pharmaceutical companies want to use QCs to simulate complex molecules and accelerate drug discovery. Big companies engaged in developing advanced batteries, especially for EVs, want to use QCs to simulate battery chemistry, which could lead to electric cars with higher mileage per recharge.

Quantum entanglement underpins quantum computing technology. Pairs of calcium monofluoride molecules have been shown to interact under optical tweezers and become entangled—a crucial effect for quantum computing. This strategy presents a promising new platform for the realization of different quantum technologies, such as computation and sensing, among other applications [15, 16].

Quantum technology (QT) is the future, and quantum computing education is linked with STEM education. All the students learning quantum computing may not get jobs in the quantum technology industry, but may instead work in a related STEM field industry, such as, fiber optics or cybersecurity, that directly benefits from QT knowledge.

4.6.9 Biocomputers

AI systems rely on a web of neural networks, in a way similar to how the human brain functions. Integrating laboratory-grown human brain tissue with conventional electronic circuits, researchers have made a hybrid biocomputer, that can handle tasks like voice recognition. The idea is to make the biological neural network within the brain organoid do computing, after building a bridge between AI and organoids, and initiate the formation of improved models of the brain in neuroscience research. Organoids, made from stem cells, can be morphed into neurons, like those we have in our brains.

4.6.10 Mapping of icebergs

The polar environment is ever-changing. Icebergs, sometimes very large, drift through icy waters, presenting hazards for both maritime navigation and the marine ecosystem as they melt. AI allows for near-instantaneous mapping of icebergs, crucial for maritime safety. In one-hundredth of a second, the AI can delineate the surface area and outline of icebergs with meticulous precision. Human analysis is time-consuming and laborious. Neural networks are used to 'train' a computer to accurately map the outline of icebergs from images taken by satellites [17].

4.6.11 Jobs at LHC and at fusion tokamaks like ITER

The Large Hadron Collider (LHC) announced the discovery of the Higgs boson (the particles which gives mass to other particles) in 2012. Physicists have proposed several designs for accelerators that would produce vast numbers of Higgs bosons and enable precise measurements of their interactions with other particles, to test the standard model, or to propose a new theory about the origin of the universe and evolution. Study of dark matter, neutrinos and the Higgs boson are all tied up with big accelerators or colliders, where engineers and scientists from different subjects need to work together, implying job opportunities for STEM PhDs.

STEM PhDs can also aspire to find employment at the world's largest nuclear fusion reactor, ITER (International Thermonuclear Experimental Reactor) or several smaller tokamaks being operated across the world with the aim to produce power using nuclear fusion.

Nuclear fusion energy is a future source of sustainable energy to complement renewables. Nuclear fusion occurs when deuterium (^{2}H) and tritium (^{3}H), the two isotopes of hydrogen (^{1}H), merge to release enormous energy, at a temperature of 150 million °C.

ITER is being built over the last two decades at Cadarache in France at a cost of $20 billion. The fuel in it, in the plasma state, will be held in magnetic confinement, for which over 87 000 km of thin wire of Nb–Sn has been used to fabricate 19 toroidal magnetic field coils, each weighing about 360 tons. Together they will produce a magnetic field of 11.8 T to confine the plasma.

The ITER project is an international collaboration and technological innovation, involving over 30 countries and numerous high-tech companies. Completion of the construction of the main ITER reactor and its first plasma are planned to be around 2035. When in operation, the final version of the ITER fusion reactor will generate 500 MW of output power (into the grid) on an input of just 50 MW, which may happen only around mid-2060s.

4.6.12 EVs

Developing technologies like electric vehicles laden with advanced microchips and software is turning the auto industry into a dynamic industry. That is where STEM PhDs will find job opportunities.

Many electric vehicles cost more to operate than their gasoline powered counterparts and EVs may cost more over lifetime than gas powered counterparts. However, EVs are more competitive in cities with high gasoline prices, low electricity prices, moderate climates, and direct purchase incentives, and for users with home charging access, time-of-use electricity pricing, and high annual mileage.

Freezing temperatures drain batteries and reduce driving range. Vehicles use more energy to heat their batteries and cabin in cold weather, raising the energy consumption. Tesla advises drivers to keep the charge level above 20% to soften the impact of freezing temperatures. STEM PhDs stand to gain employment opportunities in the electric car sector.

i7, an electric incarnation of a BMW looks almost identical from the outside to its internal combustion counterpart, it is quiet even at highway speeds. The car comes with a large video screen that folds down from the ceiling. BMW buys most of its batteries from suppliers in China, as does Tesla, but is also developing its own battery technology and is testing new battery designs and manufacturing processes. BMW is testing continuous flow slurry-making of lithium and other ingredients unlike in batches, as done by most manufacturers. The continuous process is faster and cheaper. On the other hand, Tesla, based in Silicon Valley, is already a leader in software and battery technology.

4.7 Societal benefits: improving the quality of life for masses

4.7.1 Climate equity specialist—a job

Climate inequity is pushing marginalized communities to be displaced owing to extreme climate disasters. This has led to the creation of a new job title, a Climate Equity Specialist, who would advise low-income families to adopt technologies ranging from solar panels to fuel-efficient tires; encourage them to use energy-efficient equipment such as compressed-air systems, freezers, and transformers; and work with local governments to garner support for these activities, as well as convince the latter to support only climate-responsive building activities [18].

4.7.2 Teaching science at school

Khan Lab School students are trying out experimental conversational chatbots that aim to simulate one-on-one human tutoring. These chatbots are based on AI models underlying chatbots like ChatGPT, and have been specifically designed for school use and respond to students in clear, smooth sentences. These automated study aids could usher in a profound shift in classroom teaching and learning. These chatbots not only extend student access to individualized tutoring, but can also help teachers with lesson planning, freeing them up to spend more time with their students.

Doing a STEM PhD is hard work, physically. Teaching science in school is also like that. During their PhD days, the STEM PhDs learn communication and presentation skills which are transferable, and serve them well to prepare them as science teachers for the school classroom. Scientists who choose to teach after their PhDs help open up science as a career path to secondary-school students.

4.8 Conclusions

Interdisciplinary fields, such as, human-PC cooperation and logical AI are developing as a result of the expansion of AI innovation. The ascent of AI consciousness involves people who can hold AI aspects in addition to human-centeredness all the while. These experts need to foster AI frameworks that are advantageous as well as straightforward.

In addition, we have not yet understood the full capacities and limits of AI's recent technological innovations. Identifying precisely the relation of science to technology is necessary to gain such an understanding and therefore to control and regulate these admittedly overwhelming advances. The tech outage on 20 July 2024 caused by a flawed update from CrowdStrike, whose software is used around the world, forced several airlines to ground flights, canceling thousands of them, across the world. Some airlines could normalize operations in 2–3 days, but some others took more time.

Using generative AI is exciting in many ways, but it also hampers faculty prospects for convincing students that writing is a cognitive tool, and precious form of human expression. Because of AI's incredible skills as author and editor, we have reached a tipping point. As academics, we need to muster the will to refocus the conversation about writing.

In science, experimentation and hypothesis generation often form an iterative cycle: a researcher asks a question, collects data, and adjusts the question or asks a fresh one. AI systems that generate hypotheses often rely on machine learning, which usually requires a lot of data. Making more papers and datasets openly available would help, but scientists also need to build AI that doesn't just operate by matching patterns but can also reason about the physical world. AI systems should not be driven only by data—they should also be guided by known laws. That way, we can include scientific knowledge into AI systems.

Finally, to answer the obvious question: why not do a PhD in AI itself to get a high-salary job? The answer is that doing a PhD in AI may indeed be tough considering the current fast-moving scenario in the development of AI.

References

[1] *Artificial Intelligence, a New Age of Possibilities, TIME (Special Edition, 84 pages)* 2024
[2] Weise K 2023 Amazon introduces Q, an A.I. chatbot for companies *New York Times* https://nytimes.com/2023/11/28/technology/amazon-ai-chatbot-q.html
[3] OECD 2023 *OECD Employment Outlook 2023: Artificial Intelligence and the Labour Market* https://oecd.org/content/dam/oecd/en/publications/reports/2023/07/oecd-employment-outlook-2023_904bcef3/08785bba-en.pdf
[4] 2024 Job title of the future—AI prompt engineer *MIT Technology Review* https://technologyreview.com/2024/04/24/1091125/ai-prompt-engineer-generative-ai-job-titles/
[5] Lu Y 2023 Generative A.I. can add $4.4 trillion in value to global economy, study says *New York Times* https://nytimes.com/2023/06/14/technology/generative-ai-global-economy.html
[6] Sanchis T 2023 Strategies for multidisciplinary research *Nat. Phys.* **19** 1736–7

[7] Lane B 2023 The die is cast: petrol and diesel engines are dying. The electric age is inevitable *Guardian* https://theguardian.com/commentisfree/2023/jun/07/petrol-diesel-engines-technology-electric-cars

[8] Fleischmann J, Hanicke M, Horetsky E, Ibrahim D, Jautelat S, Linder M, Schaufuss P, Torscht L and van de Rijt A 2023 McKinsey—Battery 2030: resilient, sustainable, and circular *Battery News* https://batteriesnews.com/mckinsey-battery-2030-resilient-sustainable-circular/

[9] MIT Open Learning 2024 MIT can give you 'superpowers' *Medium* https://medium.com/open-learning/mit-can-give-you-superpowers-1274e375a9dc

[10] Clay D E, Brugler S and Joshi B 2024 Will artificial intelligence and machine learning change agriculture: a special issue *Agron. J.* **116** 791–4

[11] Morrison S M 2023 Predicting new mineral occurrences and planetary analog environments via mineral association analysis *PNAS Nexus* **2** 1–13

[12] Banik M *et al* 2024 Delivering DNA aptamers across the blood–brain barrier reveals heterogeneous decreased ATP in different brain regions of Alzheimer's disease mouse models *ACS Cent. Sci.* **10** 1585–93

[13] Ortiz-Catalan M *et al* 2023 A highly integrated bionic hand with neural control and feedback for use in daily life *Sci. Robot.* **8** eadf7360

[14] Satterlee K 2023 The $1 cure: how programmable bacteria are reshaping cancer therapy *SciTechDaily* https://scitechdaily.com/the-1-cure-how-programmable-bacteria-are-reshaping-cancer-therapy/

[15] Holland C M, Lu Y and Cheuk L W 2023 On-demand entanglement of molecules in a reconfigurable optical tweezer array *Science* **382** 1143–7

[16] Bao Y 2023 Dipolar spin-exchange and entanglement between molecules in an optical tweezer array *Science* **382** 1138–43

[17] Braakmann-Folgmann A *et al* 2023 Mapping the extent of giant Antarctic icebergs with deep learning *Cryosphere* **17** 4675–90

[18] Chaudhary A 2024 Job title of the future: climate equity specialist *MIT Technology Review* https://technologyreview.com/2024/02/28/1088244/climate-equity-specialist-justice-energy-solutions/

Chapter 5

Mentoring, innovations, patents, entrepreneurships, and jobs

STEM students at graduate level, and more so at post-graduate level and doctoral student level, are taught and trained to test every way there is to solve a problem, until they find the one that works. By contrast humanities grads usually have a point-of-view that has come to them from their learning. Trying to explain an alternative point-of-view to a humanities student would not be as easy as to a STEM student.

All students of STEM know that at the root of all new innovations is the application of principles of science to explain observations made during the pursuit of knowledge. Science is the pursuit of seeking facts, by learning through reasoning.

While doing science we essentially try to ask basic questions, such as *'what is a God particle?'* (the name given to the Higgs Boson), which may have profound meanings. To answer such a question, we make observations. Careful observations lead to patterns; patterns lead to further questions! A 'good' question can lead to new discoveries; though not all of them, but it is common knowledge that some key discoveries can lead to patents, innovations, prototypes, and finally useful commercial products and/or new sustainable technologies for health, medicine, environment, and energy sources for good living conditions for all humans, globally, and in the process generate new job opportunities.

In this chapter we describe how, under appropriate mentoring and direction, a fresh PhD in STEM, who might be working as a postdoc perhaps, can discover innovative ideas or new materials and patent them. That inspires them to go for entrepreneurship, perhaps in the form of a new start-up, and demonstrate a prototype device based on their ideas/patent and then commercialize it into a successful product acceptable to the industry. This sequence can facilitate employment opportunities for them and many other colleagues ready to work with them. Some of the viewpoints described in this chapter also complement the discussion provided on the generation of new jobs in other chapters of this book.

5.1 Introduction

Automation has historically displaced human workers in factories (e.g., automotive manufacturing) or when performing routine computational tasks. Will generative artificial intelligence (AI) tools such as ChatGPT disrupt the labor market by making educated professionals obsolete, or will these tools complement their skills and enhance productivity?

The sudden emergence of ChatGPT in November 2022 has emphasized that twists and turns may lie ahead, but a STEM PhD would begin their career as before, nonetheless. They may need to re-skill to move forward towards their future. Manufacturing companies need to address the impact of digitization and automation in their production lines. At times, employers may have to consider coordinating the skills of both the manufacturing workers and those of research scientists working in their R&D labs.

Graduate and even post-graduate degree holders appear to be rapidly becoming redundant with the arrival of AI. With the advent of AI and with robots already doing repetitive tasks for us, human creativity may be valued more, and being able to 'think' and daydream like a human might become valuable, as opposed to the 'hard work' of memorizing formulae.

With AI coming at us at a fast pace, the only logical way may be to co-operate with AI and move forward, learning where human inputs are crucial. After all, AI won't install your solar panels or cut your hair. The largest automobile company in the world, Toyota, has been adding robots at its assembly line for the past several years, but it also ensured that its employees were not made redundant and were retained and allotted other jobs like a supervisory role requiring a human insight on the shop floor.

Community life on campus offers an opportunity to students to develop skills other than those taught in class, and to get involved in accordance with their convictions while socializing with their fellow-students. The campus is also seen as the place where graduates can interact with working professionals, as a space for learning, discussions, and a place where students can hone their skills on subjects like environment, ethics, diversity, and inclusion.

Keeping the theme of this book in focus, we need to reposition our understanding of 'employability' by considering the uncertainties about what AI will mean for the careers of STEM PhDs. Covid-19 has taught us that future graduates will need to be highly flexible in order to tackle the rapid social changes we are undergoing. The humanities, social science, general science, technology, and creative industries sectors such as design, can deliver adaptable, flexible mindsets. These generalist graduates do possess some basic and useful soft skills such as emotional intelligence, communication, and teamwork.

Specialized technical courses designed for, say, rocket scientists, who are not generalists would require some 'generalist' or 'humanities' content. Likewise, the generalist courses should have some technical content. Administrators, for instance, which is where most generalist degree holders finish up, will be better off with some knowledge of the basic technical details of what they are administrating. This should be viewed in the context of meeting the requirements of employers who are increasingly looking for people with analytical and problem-solving skills.

Before we move on from the subject of the 'employability' of STEM PhD students, note that, while scouting for jobs, there are opportunities for them to find them at job fairs, which provide the candidates with a chance to discuss their capabilities with people from the industries offering jobs. These in-person meetings sometimes lead to job appointments beyond what a résumé posted online would have resulted in. This is because job-seekers can communicate with prospective employers face to face, and try to convince them about their suitability for the job being offered. That is where the body language of the candidate helps, and may result in a short impromptu interview on the spot. Further, it is a good idea, in any case, to follow up by sending an email and an updated CV to prospective employers that the candidate might have met at the job fair.

Big technical companies with available jobs advertise for applications from eligible candidates for their summer internship program but they may eventually select and hire only a small number of interns. In several places in this book we shall discuss the opportunities and benefits of being selected as an intern in a big company towards landing a regular job in the same company.

5.2 What is an innovation?

An innovation is the transformation of an idea or invention into goods/services that create value or for which customers will pay. To be called an innovation, an idea must be economically replicable and must satisfy a specific need.

Innovation is the introduction of something new, not just inventing something new, but a product that can be launched and 'introduced' to the world. This definition of innovation covers not just new products and services, but also improvements in business processes or business models.

Innovations and new technologies generate immense job opportunities, as happened during and after the industrial revolution. However, some recent new digital innovations/technologies, such as ChatGPT, AI, IoT, robotics, machine learning, 3D printing, and 5G, etc, though exciting, could also swallow some existing jobs, even as they create opportunities for completely new jobs. A clear indicator of this is IT—possibly the largest private sector employer in recent times, which is not growing now as it was before 2011, and a large percentage of traditional IT jobs are predicted to be lost to automation in the coming years.

Traditionally, innovation has meant that basic research leads to new inventions, which are then patented (if possible) and developed into commercial applications. But the internet has changed the whole scenario. Information, knowledge, cutting-edge tools, and software are now available almost instantly and at a fraction of the previous cost in virtually any industry, and most of the innovations do not rely on commercialization of basic research, but deal with applying existing knowledge, technologies, and resources to solve given business problems.

Innovation is a team effort. The myth of the lonely inventor, the solitary genius, is gone because innovation currently requires collaboration and sharing.

Innovation happens in places where people do not live in isolation, and are free to think, experiment, and speculate. And where they can trade with each other and exchange ideas frequently [1].

Hence, even though most businesses don't have the resources to compete with top universities and large corporations in foundational research, they can still have many opportunities for innovating incrementally around customer service or by automating internal processes. Thus, there are typically plenty of opportunities where even minor adjustments could lead to significant savings in cost or time. The beauty of addressing these low-hanging fruits is that the benefits start to flow in quickly, which leads to a significant impact in the long run. In the last few years, we've seen more and more industries adopt innovative business models, such as moving from selling individual products to subscription services and value-based pricing models.

In America, you are now at least 700 times more likely to die in a car, per mile traveled, than in a plane. The decline in air accidents is as steep and impressive as the decline in the cost of microchips as a result of Moore's Law. How has this been achieved? The answer, as with most innovations, is that it happened incrementally as a result of many different people trying many different things.

For society at large, innovation is a key driver of economic growth. In an industry, such as online search and computer operating systems, a few companies ride a wave of innovation, achieve a market leadership and often undercut competition, driving it out.

The impact of radio on the polarization of society was huge—shades of what has happened more recently with social media. The first railway in America began operating in 1828, in France in 1830, in Belgium and Germany in 1835, in Canada in 1836, in India, Cuba, and Russia in 1837, in the Netherlands in 1839.

It was synthetic fertilizer that enabled Europe, the Americas, China, and India to escape mass starvation and consign famine largely to the history books. The Green Revolution of the 1960s and 1970s was about new varieties of crops, the key to which was that they could absorb more nitrogen and yield more food.

5.2.1 Innovations can be risk-prone

While innovation is risky, not innovating is the biggest risk of all. Without innovation, you are guaranteed to go out of business, sooner rather than later.

The Jersey City case proved to be a turning point. Cities all over the world began using chlorination to clean up water supplies, as they do to this day. Typhoid, cholera, and diarrhea epidemics rapidly disappeared.

The development of civil nuclear power was a triumph of applied science, the trail leading from the discovery of nuclear fission and the chain reaction through the Manhattan Project's conversion of a theory into a bomb, to the gradual engineering of a controlled nuclear fission reaction and its application to boiling water. In terms of its energy density, nuclear power is without equal: an object the size of a suitcase, suitably plumbed in, can power a town or an aircraft carrier almost indefinitely.

However, following the Three-Mile Island accident in 1979, and Chernobyl in 1986, activists and the public demanded greater safety standards. They got them. According to one estimate, for the same unit of power produced, coal kills nearly 2000 times as many people as nuclear, bioenergy 50 times, gas 40 times, hydro 15 times, solar five times (counting risks during rooftop panel installation), and even wind power kills nearly twice as many as nuclear.

5.2.2 Encouraging innovations

There is a 'quiet crisis' in the development of the talent pool. The current system of education fails to draw young women and minority students into fields such as computer science, engineering, physics, and mathematics. Failing to inspire them to participate in STEM leaves our innovation system wanting.

There is no innovation without innovators. Therefore, the world needs to invest more in human capital. An example is the US innovation ecosystem strengthened by US universities which continue to draw the brightest students in science and engineering from around the world.

The breeding ground for innovation in a country can be set out successfully only if it is supported by a vibrant higher education sector, a strong focus on research and an entrepreneurial spirit within academia. Universities should allow inventors to hold a majority share of the patent, and a major share of the royalty for their own inventions. Reaching out to industry by them should be facilitated for efficient technology transfer.

Advances in technology can lead to a slow-down of the influence of a potent innovation. Nuance Communications, Inc., a computer software technology corporation introduced a 'large vocabulary continuous speech recognition system', in the late 2000s. But soon it was not the only company to realize that voice was poised to become a prime channel for human interaction with computers and cloud devices. And voice recognition wasn't just about dictating text, but emerged as equally or even more useful for shopping, searching for information, selecting music and video entertainment, and controlling appliances. Sans keyboard and a mouse, it was truly revolutionary. Siri was based on Nuance-technology. Alexa came along and Amazon has a large number of engineers working on Alexa products. Nuance couldn't compete and was acquired by Microsoft in 2021.

In 2018, Canada established five innovation 'superclusters'—industry-led, public–private collaborations located across Canada to work on areas like artificial intelligence, ocean-based technologies, etc, which were awarded Can$750 million in the fiscal year 2022 budget to continue their work for another six years [2]. The Canadian government announced that it will invest Can$1 billion (about US$780 million) over the next five years to create a funding agency focused on innovation in science and technology. A Can$15-billion Canada Growth Fund aims to stimulate private investment in low-carbon industries and restructuring supply chains.

5.3 Patents

Patents are supposed to protect and promote innovation. To secure a patent, an invention must be truly novel and non-obvious, it must be described in enough detail for a reasonably qualified person to build and use it, and it must actually work.

To get a patent, the invention must be new, not published in any form before, including as an academic publication by the inventors themselves. Delivering a talk at a scientific conference or demonstrating the idea can kill a patent application.

There must be an element of non-obvious inventiveness in the idea of a patent. Also, the disclosure in a patent must be sufficient for a skilled person to reproduce the invention with usual effort.

Before a company can start selling a product, it must protect its IP (intellectual property) by patenting the technology that makes it special. To pre-empt a discovery being scooped, one may publish a peer-reviewed paper on it. However, considering the commercial angle, the patent filing should come first, in order to disclose in it a fresh invention.

The main objective of a patent is to ensure that a start-up company can secure investment for technical development. International treaties usually allow a patent application in one country to establish priority in the rest of the world.

The share of innovations that are patented varies across industries. In manufacturing machinery, roughly 50% of all innovations are patented. In chemistry hardly anything is patented.

Most popular drugs are protected by tens of patents each, though those patents have little by way of new addition. When it comes to protecting a drug monopoly, no modification is too small—making a tablet instead of a capsule, changing the dose, adding a flavor, and so on.

Timing of patenting is crucial. In 1876, in what is often cited as a spectacular case of simultaneous invention, Alexander Graham Bell arrived at the patent office to file a patent on the invention of the telephone, and just two hours later Elisha Gray arrived at the same patent office with an application for the very same thing.

If you had developed an improved dye in the 1850s, there was not much advantage patenting it because you had to disclose your innovation, and then the patent would expire, so it was better to just keep it secret.

Thomas Edison invented the light bulb. But so did Marcellin Jobard in Belgium; and so did a few other inventors from England, Russia, Germany, France, Canada, and America, in separate efforts. Every single one of these people produced, published, or patented the idea of a glowing filament in a bulb of glass. But Edison got credit, because he set up a laboratory in Menlo Park, New Jersey, to do what he called 'the invention business'.

Patents have steep costs. They slow down the wheels of innovation by making it harder for would-be inventors to proceed with their work. They strain budgets by preventing cheaper products from entering the market. And they leave honest inventors vulnerable to patent trolls—people who buy up weak patents not to create anything new or useful but to hold legitimate inventions to ransom.

The best solution to ensure that patents spur innovation instead of thwarting it is to set a high standard for what deserves patent protection. And, make it easier to challenge bad patents before they are granted. The patent office could also develop a sliding scale system in which the largest and wealthiest patent filers subsidize the smallest and least wealthy [3].

Innovation and patent activity are related, but innovation is a process, and patents are the potential outcomes of that process. In a bid to go beyond simply counting the patents, STEM PhD students need to develop the ideation culture that produced those patents, and then follow it up to commercialization.

5.3.1 IP rules, copyrights, and patents

A patent is an intellectual property right to provide natural protection against competition.

In the 1980s, US-based pharmaceutical companies led a push to expand IP rights globally at a time when the firms faced potential competition from a growing generic-drug industry in India and Brazil, even though such restrictive policies were sure to limit access to life-saving medicines. The TRIPS Agreement was passed in 1995, implying that over 150 countries belonging to the World Trade Organization (WTO) must adhere to IP rules, such as respecting patents for a minimum of 20 years.

Next-generation mRNA vaccines need to be developed without the entanglement of large pharma patents.

Innovations in content—books, music, and so on—are very important to our society, but are protected by copyrights, not patents.

Many research institutes care more about publications than patents and have limited funds for filing patents. Some institutes evaluate invention disclosures for patentability and commercial success before deciding to file a patent.

5.4 Prototypes

Unlike the scholar, who is captivated by novel science and discovery, those in industry have up-front concerns about replicability, with a specific knowledge that the findings can be replicated at scale, using commonly-known assays under standardized conditions. Qualified scientists with exposure to industry can help faculty do research that meets the industrial requirements.

Typically, an academic scientist's concept for a drug or vaccine is based on results in a cell or a test tube involving a biological mechanism that may not be fully appreciated by an outside audience, even biotech experts. The ultimate demonstration of robustness is a prototype. Prototyping builds the idea into a stripped-down version of a device (or therapy) that can answer routine questions by people from medicine (how long does a dose last?, can it be ingested rather than injected?, what side effects does it cause?, etc). Then there are questions like does it have a positive effect on the targeted area in an animal model?

5.5 Entrepreneurship

Entrepreneurship is tough. Having your mentors with you gives you renewed motivation to try and aim to become an entrepreneur. For this, mentors need to continue to meet with entrepreneurs regularly as they proceed with their business. Without an identified potential mentor, students are not able to benefit adequately from workshops, skill up-grade sessions, or counseling for careers.

5.5.1 Mentoring for entrepreneurship

Aspiring researcher entrepreneurs are often asked to show patents to demonstrate the viability of their technological innovation to secure external funding. Mentorship is useful for researchers to transform their ideas into patents and businesses.

For many doctoral students, cultivating productive mentor relationships is a greater and more pressing concern than landing a sustainable postdoctoral career path. This is because in the early stages of their doctoral program, many of them can't even envision continuing or completing it, so they don't invest much in anything that pushes it forward. Subsequently, they get busy with advisers, their research studies, presentations, and writing of manuscripts and dissertation, leaving hardly any time for mentorship. Most mentorships begin towards the end of doctoral program, to facilitate their entry into a career.

A good mentor can have a huge impact on your life by teaching you how to navigate the challenges you never saw coming. When you ask someone to be your mentor, you are also asking them to invest their time in you. Show them that their time is being well-used by demonstrating a return on their investment by checking if you can help them in any way.

STEM students have concerns related to mentoring in lab settings. They often assume that a 'mentor' and a 'PhD program adviser' are the same. They need not be, because a mentor does not need to be a faculty member in our doctoral program, or even a faculty member at all. On the positive side, some students are drawn to a course because they had benefitted from effective and committed mentorship.

Students and faculty members often work in a climate in a university which is not hospitable to good mentorship. One reason for that is that their advisers had no training in formal mentorship. Often part of the mentorship is undertaken by a 'ghost adviser' who may be a junior faculty member, or a senior doctoral student.

5.5.2 How do entrepreneurs succeed in their pitches?

Young entrepreneurs looking for finance need to pitch their project to investors. Entrepreneurial pitches incorporate verbal and non-verbal communication allowing the entrepreneurs to present themselves to investors as legitimate people to develop their ideas. The idea is to increase the willingness of investors to unlock funds.

An ideal pitch can be prepared based on a mastered self-presentation, adapted for the stage.

Investors prefer humble entrepreneurs who are open to listening to suggestions and a strong disposition to follow the advice of their partners. An appeal to the emotions has a role in the success of a pitch in order to convince the potential investors about the project. Minor exaggeration during the presentation of the pitch is fine and is perceived as legitimate as long as the entrepreneur does not drift into distortion of the facts. Gestures and facial expression of emotions can positively influence investor reaction. A metaphorical gesture or a smile, especially at the beginning and end of the pitch, accentuates the quality of the presentation and is likely to influence the decision of investors, positively. The frequency and duration of smiles must be controlled, lest the smiles are taken to presume that the entrepreneur lacks seriousness.

5.5.3 Spreading entrepreneurial literacy

The Shift Retail Lab is a student retail lab under The Virginia Commonwealth University (VCU)'s da Vinci Center for Innovation, which has run a master's degree program in product innovation, in the USA, since 2012. VCU has been recognized for its economic and social impact through innovation and entrepreneurship, technology transfer, talent and workforce development, and community development.

VCU's Entrepreneurship Academy brought together 150 first-generation and low-income students alongside 50 community members identified by community partners to develop learning modules in topics such as design thinking, digital literacy, and the art of the pitch [4].

5.6 Ideas for entrepreneurships and start-ups by STEM PhDs

STEM PhD students possess brilliant brains, fine-tuned to do research. A career for them should be within the scope of their skill-set. The trick is to treat the job hunt like a research problem. Making an amazing discovery is not enough for PhD degree holder scientists, they need to publish it so that the scientific community knows about it.

Similarly, in business, one may have an idea that will provide a solution to a certain problem, but there must be customers who are willing to pay for that solution. One needs to propagate this among prospective customers using social media, paid advertising, and emails for marketing. But to scale up and professionalize a business, investments are needed and risk-taking is also necessary. That's why most business experts recommend starting this career path as a side job, or having a cash reserve equivalent to at least a full year's living expenses.

Entrepreneurship among university researchers is fostered by encouraging researchers to establish their own start-ups. A robust supply chain is essential to bring out full technology transfer potential of the entrepreneurs. In view of spin-offs' limited cash flow, universities can accept shares in lieu of licensing fees and royalties.

For a start-up, a good strategy is to start a project to tackle a problem looking for a solution. It is advisable to simultaneously work for a PhD program and run a start-up to save time, effort, and priority.

Networking is another key activity that helps students to build creativity and innovation by working on project-oriented courses, working at student incubators, and obtaining product development assistance, and internships with start-ups. The idea is to get hands-on experience to solve real-world problems. This includes working on advanced battery research, or developing new vaccines for health-care, and so on.

Even as the college-to-job landscape is expanding with increased interest in internships, mentoring, and experiential learning, the majority of students at different colleges get little attention and resources for such activities.

One can't take the 'obvious' route to solve problems, and that is where the role of a STEM PhD would be vital. For instance, a very high rate of childhood pneumonia in Bangladesh was initially thought obviously to be due to pathogens, but it turned out after a deep study that it was largely due to air pollution by black carbon particulates that spew from the country's thousands of unregulated kilns used in brick manufacturing. But the brick-manufacturers couldn't afford to install scrubbers costing $5000 each. A simple solution was devised to stack unbaked bricks in a configuration that improves airflow, increases the insulation by the ash—resulting in less coal use by the kiln, reducing pollution, and producing higher-quality bricks. In fact, it serves to achieve one among many sustainable development goals (SDGs).

Dhaka muslin, a thin cotton cloth was famous for centuries, until it disappeared after the East India Co came to India. They destroyed the culture of weaving thread counts in the range of 800–1200 meaning that they contain roughly that number of horizontal and vertical threads per square inch of fabric, an order of magnitude above any other cotton fabric made today which have thread counts only between 40 and 80. The secret was the special cotton grown on the banks of Meghna river used to make it, and the innovating artisans who were weaving it [5].

We need companies that will bring us better batteries, solar panels, windmills, electrical conduction, maglev trains, and AI. Emphasis on wearable tech and early illness detection will definitely grow in view of the sustainable health programs across the world.

There are many ideas, but starting a company and running a company are two different skills; one is innovation, and the other is book keeping. You can't teach the former, but Business Schools teach the latter. With the confidence gained in bringing up innovative ideas, STEM PhDs are ideally placed to solve the problems related to SDGs. In this context, we describe below some potential problems that the STEM PhDs could try to work on solving, while generating new opportunities for jobs as well:

(i) The oscillating lift and drag forces generated by each wing of a hovering hummingbird radiate the distinctive humming timbre. A study of this could feature in the development of improved drones by guiding bio-inspired engineers how to design more silent flapping robots.

(ii) The integration of flexible electronics and biotissues is a transformative technology that advances the fields of health-care, human-machine interfaces, and IoT by enabling devices with the capability to interact with the human body. However, living systems and electronics are inherently different. Biotissues are wet and soft, while electronics are dry, relatively stiff, and non-bioactive; biotissues conduct ions, while electronic materials often conduct electrons. The mismatch between these two categories of materials can lead to unstable bioelectronic interfaces between the bioenvironments of the human body. One needs to form a stable interface for bidirectional electrical communications between electronics and the living systems, which is difficult [6].

(iii) Hydrogels possess high biocompatibility and tissue-like mechanical properties, and thus are promising in tissue engineering. Hydrogels have been successfully implanted in the body to bridge electronics and biotissues. Conducting hydrogels can form an electrical bio-adhesive interface, obviating the need for suturing or physical attachment.

(iv) There is scope for a start-up to work on a microfluidic chip that can channel the flow of a tiny amount of fluid for early cancer screening by detecting cancer biomarkers.

(v) There are a growing number of 'green boom-towns' around the world using the transition away from fossil fuels to boost growth and stop the economic and demographic declines triggered by off-shoring. It also generates jobs for the local workforce, which is drawn from the battery, oil and gas, pharmaceutical, paper, and chemical industries. More than $100 billion in investment for green steel, railways, ports, renewables, and ammonia plants will soon start flowing into northern Sweden. Locations have been identified as future sites for hydrogen-based steel plants, which are expected to consume a large percentage of the nation's green power in the coming decades [7].

5.7 Jobs

5.7.1 Career guidance

Career advice for students is hardly done, and at best it is an optional service. Career readiness should be treated as a core education component that is embedded in college education. Students should be told how the courses they study after leaving school might affect their employment prospects.

Industries such as education believe that employees work for something other than money. 'You don't do teaching for the money, you do it because you believe in the kids that come into your classrooms.' But this discourages STEM PhDs to go for education jobs.

The gender gap needs to be closed. More women should be encouraged to take up apprenticeships in STEM. A lack of science teachers in schools and dwindling apprenticeship opportunities are denying women opportunities to build STEM careers of the future. MotherCoders was started to recognize that organizations that offer coding opportunities for girls don't always support mothers. That includes options such as evening classes, online learning and weekend boot-camps for people with care-giving responsibilities. Similarly, we have R-Ladies and Girls Who Code. Now we need attention for mothers with STEM qualifications.

5.7.2 Jobs for STEM PhDs

5.7.2.1 Postdoc

Doing a postdoc is often seen as a valuable period when a PhD degree holder can concentrate on their intellectual development. But a question is often asked: 'why do a postdoc?'. The postdoc is often necessary to eventually obtain a position for which the requirements are multiple: to have conducted quality doctoral research, to have published results from your thesis or to show already well-developed publication projects, have taught a sufficient number of hours with a variety of audiences, be integrated into the networks of their scientific community, be involved in the activities of their department and/or their team, etc.

For a large number of PhD students who do not wish to move away from academic research and/or teaching, an obvious choice after completion of their PhD is doing postdoc as an immediate career option. A postdoc is a useful option for working on a STEM project while earning a stipend. Although it is a temporary position, it provides ample opportunities for refining one's teaching and research skills, thereby can be treated as a stepping stone, even if temporary, to grow in one's career and steer it towards a permanent position in academic research, teaching faculty, or even in industry.

The job of a postdoc often entails activities related to: both collaborative and personal research; laboratory services; project management; organization of scientific events; and, teaching. This places the postdoctoral fellow in the position of a teacher–researcher, ready to start their first academic position. It is a good balance if postdocs can spend about 50% of their time doing personal research, which is essential to push their post-thesis projects and find a permanent position.

One highlight of the postdoc is to move from their original laboratory, from their thesis supervisor, to affirm their project, and ultimately, to contribute to building their academic identity. The postdoc makes it possible to develop their own scientific identity and to get out of what is sometimes considered as the shadow of the thesis supervisor.

Post-doctoral fellows (PDFs) can be funded in three ways: (i) the PDF secures funding themself; (ii) a principle investigator secures a research grant that includes funding of postdocs; and (iii) a large research body hires postdocs to work for a project.

Generally, part of a research grant secured for a project is used for hiring of one or more PDFs, with the condition that the PDF will then work exclusively for that project.

Sometimes, under certain exclusive programs, postdocs can secure funding themselves, in which case they are offered a stipend and they can then take the funding of the research project to any lab/university of their choice. An example is the Marie Sklodowska–Curie Fellowship which can be worth over GBP 50 000 per year.

Most PDFs stay as postdocs for an average of 2–4 years before moving on to a regular faculty position or another fellowship. This allows them to contribute significantly to a project, publish a few papers, and attend some conferences. A stay of 2–4 years or even beyond helps PDFs to interact deeply with the students they teach or supervise. Doing a PDF is beneficial for early-career researchers, since they can keep trying for a regular and stable position, while staying active. This is in line with the advice given often to young researchers to work as hard as possible and strive to be the best at what they are doing, and let the money take care of itself, in order that they can achieve the goal of getting their dream career.

5.7.2.2 Successful transition from postdoctoral position into a faculty job
In today's faculty job market, some candidates are oriented toward research and others toward teaching, some are on the tenure-track and some are not. In recruiting

for tenure-track jobs, some universities are looking almost exclusively for a candidate to teach, while others also want someone who can direct discipline-based education research or fill a leadership role in the department. Whatever the position's particular mix of teaching, research, and service may be, the priority in hiring is almost always the department's teaching needs.

There is always a lot to learn for a postdoc in any specialist field, despite heavy teaching responsibilities and time-consuming administrative tasks. There are few options for PhDs outside of academia, because they are considered so specialized that many academics—particularly those in the humanities—can't imagine what else they might do. The arts are in decline and offer decreasing possibilities of a faculty job. Therefore, one has to look beyond the arts for a career that hasn't already been replaced by AI.

During a postdoc position, early-career reseearchers learns new skills, and add to their record of publications, thereby creating their niche. They also gain experience in applying for funding and asking for help in their job search, thus trying to establish their long-term goals. All this qualifies them to apply for a faculty position, to succeed in which the postdoc needs to prepare a research statement, and a teaching statement, in addition to getting a letter of recommendation and a cover letter from a senior faculty member, and a diversity statement from the institution where they work as postdoc. Having obtained all these papers, they finally have to prepare to make preparations for an interview, a research talk, and a 'chalk talk', too.

What is a chalk talk? Many search committees use a chalk talk to assess suitable candidates for faculty or independent research positions. The informal nature of this format (no slides, just your wits, and a board) allows them to learn about you as a potential colleague. As such, the chalk talk is frequently the most crucial determinant in landing a position.

5.7.2.3 Bridging the gap between academic research and industry
A rigorous academic researcher works very differently from the policy problem-solver. A Professor may have a likely solution to a vexing problem but would hate to get into policy, because that may derail their ongoing studies in the lab. In traditional academia, it is a regular practice to just define problems, hardly ever working on solutions needed by society. Doing that you can get a PhD, a tenure, awards, and retire never having done anything about solutions. What the world needs now is a shift in focus simply because the policy-makers don't benefit from this attitude, and remain clueless since they don't have crystal-clear ideas of academic research outputs.

It is not that doing good science is a bad thing, but most academic researchers lack strong instincts on how to use novel ideas in science for the end user. This gap has to be closed.

Bridging the gap between academic research and real-world impact means converting the laboratory results to a working device to a prototype, and then become an entrepreneur who possibly transfers the technology to a start-up or a company for commercialization of the new product. A typical example of this is when you plan to make a new drug to cure a disease, you notice that the gap between the lab bench and the clinical testing of your product and then its launching in the marketplace is huge, and the costs are also high. The worst thing is that the odds to succeed finally are generally so low that the whole thing is often called the valley of death. In fact, the time taken to take the idea of that new medicine to its first test on patients can easily take 8–10 years. Similar challenges are faced in fields other than medicine, too, where a researcher in an academic world finds themselves standing on the edge of a 'valley of death'. This is mainly because labs do not work like companies, and professors in academic institutions like universities are not entrepreneurs in the true sense.

Academics do great stuff while making the discovery of an idea or a new material which makes an impact and brings recognition, but they have not been trained to take their product through various steps to commercialization. If that gap can be bridged then the academic researchers turned say, molecular engineers can make products and solutions for climate and sustainability, using the very tools that are used to make medicines. For that, they can learn the requisite skills of expert molecular engineers who already make protein therapeutics, i.e., protein-based medicines.

A praiseworthy step in this direction has recently been taken by Stanford University by setting up at its campus what is known as Innovative Medicine Accelerator (IMA), an integrated non-profit biotech company. The scientists and staff of IMA undertake to execute workplans of yet-unmet medical requirement, which also challenge scientific innovation, and in a manner that they can develop drugs economically just by recovering the cost (which wouldn't be feasible if done by a private company), even if the project is risk-prone. The IMA's scientists put that ability within a professors' reach. Now the industrial scientist is induced to work with these people to develop a real drug [8]. The goal of IMA is not to replicate the production of drugs already available, but attempt to find solutions to discovering drugs for many diseases which are incurable, yet.

5.7.2.4 Interface of academe and industry
A PhD scholar often gets to hear during their days of PhD work that a PhD should have no trouble getting a non-faculty job because holding a PhD means that one is very bright. But the realities are just the opposite. PhDs often struggle to find a meaningful non-academic profession, because employers consider them to be lacking in 'real-world' experience for management positions.

PhD programs usually prepare scholars-in-training to become professors, and nothing else. Most doctoral students are aware that they may not be able to secure academic jobs and must resort to 'plan B', which means looking for a job in industry.

A doctorate generally closes the doors that lead to non-academic jobs in industry. PhDs don't have extensive experience in marketing or business analysis, as MBAs would have. Some PhDs have near zero work experience in statistics or coding. That restricts opportunities for getting hired for a non-academic job. Until recently, a

STEM PhD was a 'scholar', but what should describe them now? An 'analyst,' a 'consultant,' a 'project manager,' or a 'strategist'? It sounds confusing.

Non-academic employers want to see your résumé and not a typical academic CV. Besides, chances are that you were never a part of professional network in the industries and companies, which offer jobs.

PhD holders are non-traditional job applicants, hard to get noticed amongst crowds of smart undergraduates and MBAs who are also well-connected. So, PhD holders need allies!

Plan how to handle the question 'What made you quit the search for a faculty career?' In other words, try to ask the question on behalf of the prospective employer: 'Why should I employ you, when I can easily get MBAs, or graduates to do that job?'

PhD programs mostly train students in the skills they will need for a lifetime in an academic lab, such as running experiments, securing grants, and writing papers. But this does not prepare them for the careers that many PhD graduates are dreaming about, viz. in industry.

This change has pushed some universities to provide the requisite skills and experiences, by offering programs that introduce students to industry earlier in their careers—often through internships, which helps them establish ties to industry, gain new skills, and keep them motivated.

Young STEM PhDs look for new challenging jobs in technology, education, and medicine. Hence, a culture is developing amongst PhD students to split their time between coursework, research and working with an external company. More and more PhD graduates are now heading into industry or business, or even for jobs in the government sector. Around the world, increasing numbers of PhD programs are being offered that are a part of company-funded projects, providing the research students with an opportunity to work on the interface between academia and industry. Industry-based PhD and postdoc positions bridge the gap between university research and commercial products. Those who participate in these hybrid positions can gain new insights and career options beyond academia. For PhD candidates working at such hybrid projects brings interaction with future employers, whereas the companies supporting such projects can recruit talent, while expanding, simultaneously, their research capabilities.

It is easier for PhD scholars who have a working relationship with an industry to find employment, compared with PhD students working exclusively at universities. The next step is to be able to work in a university lab, while not only keeping contacts at a manufacturing company, but also with a government agency dealing with scientific research.

To summarize, there is a need to bridge the 'valley of death' between academic research and business, where promising basic research often fails to get picked up and commercialized by private companies.

5.7.2.5 Postdoc versus industry jobs
Those who take positions in the private sector, after securing a PhD, typically receive higher starting salaries than those accepting postdocs in academia. Despite

the lower pay, more of the graduates accepting a postdoc find their work challenging, and appear to be satisfied with their employment.

Both postdocs and those employed in potentially permanent positions do have it in them to solve technical problems, work in a team, and do programming. Although academics might have to accept a more junior position when they move to industry, the skills they have acquired, even as early-career researchers, mean that they could progress rapidly.

The government sector generally employs only a tenth of new PhDs with permanent positions, regardless of employment type.

Getting a call for interview for an industrial job doesn't guarantee a job. Things can take time. Therefore, postdocs are advised to start preparing for their exit long before getting the final offer of a regular job.

Infrequent contacts (weak ties) are known to provide more information on new employment opportunities, because they deliver more novel information than strong ties. This is because weak ties provide access to varied and novel information by connecting us to human social networks which are disparate and diverse. Weak ties, via social networking, e.g. LinkedIn, are thought to be specifically well suited to deliver new employment opportunities because they provide novel labor market information, making job mobility possible. Whereas weak ties deliver more job applications to high-tech industries, strong ties lead to more job applications to low-tech industries [9].

Unlike in academia, there are manifold trajectories that a person's career can take in industry.

Coming from academia, people are used to having a lot of unstructured time and if that matters to you, then a position that allows you to have a hybrid or remote-working arrangement could matter a lot to you. Others who are used to the laboratory bench might find they are more productive in an office, surrounded by people.

Some experts believe that the procedure to apply for and get a PhD degree can be split into two streams: research-focused doctorates intended for academic jobs, and practitioner doctorates aimed at industry.

5.8 Innovations by STEM PhDs towards SDGs for techno-entrepreneurships and jobs

Science, technology, and innovation are a potent strategy promoted to aid the implementation of the 17 Sustainable Development Goals (SDG) proposed by the UN. These include gene-editing and breeding technology to increase crop resistance to natural disasters and insects for improving yields; biomedical advancements, including the development of new vaccines and medicines, to reduce infectious diseases and keep people healthy; transition to renewable energy to reduce carbon emissions, improve air quality, and mitigate climate change impacts; satellite observation and spatial analysis to monitor air, water, soil, and urban development, which can help find pollution sources and changes in land cover; and digitization and communication technologies to help improve efficiencies in industrial production, transportation, and government management [10].

Innovations promote well-being. It is indeed due to innovations that people live lives of prosperity compared with their ancestors. Improvements in technology are not driven by inventors, but by innovators. Inventors advance scientific knowledge. Innovators use trial and error to make thousands of little changes to a product or a process, already established, in order to get it to work at scale.

Engineers among STEM doctorates possess special skills because they are trained to use logic, foresight and lateral thinking, and are, therefore, ideally placed to solve engineering problems with common sense.

At a formal student innovation centre, students from all disciplines work, learn, and get inspired by each other. Skills could vary from working on engine testing, to 3D printing and digital media, to culinary creations.

Too much venture capital has been going to e-commerce facilitator companies. Tech jobs are the need of the moment. Study economics and organize small farm cooperatives; study oceanography and find solutions for the acidification of the seas; study chemistry and develop new treatments for Alzheimer's; study urban planning and create walkable healthier cities. And we need innovators for clean energy, automation, medicine and infrastructure.

Laboratory insights into disease need support of specialized infrastructure to thrive, and result in quickening the drug development from bench to bedside.

According to the World Economic Forum's (WEF) report on the future of jobs, the top emerging jobs in the US are of big data architects, automation technicians and engineers, renewable energy engineers, organizational development specialists, new technology specialists, IT administrators, digital transformation specialists, IT project managers, and data analysts.

Creative, mission-driven and prestigious jobs often take advantage of an employee's passion for what they do. This includes expecting to work overtime without pay or asking people to do demeaning tasks, not in line with their job descriptions, if the workers are thought to be passionate about what they do.

5.8.1 Health sector and health-care

The challenge for science and pharma industry is to develop drugs which are successful and specific to a disease. Over the recent past, scientists tackled a totally unknown disease, Covid-19, developing new vaccines within months, which were effective against the virus that caused the disease and the pandemic.

However, we don't yet have a treatment and cure for Alzheimer's, a devastating disease, known to humankind for over a century. Alzheimer's disease can be present for many years, before symptoms appear. One can't give a toxic long-term drug to someone who may or may not develop the disease, after several years. Research into it is severely under-funded as compared to cancer, HIV/AIDS, or Covid-19. Alzheimer's is the most complex disease of the brain, and the human brain is a very complex organ.

As a potential problem to be tackled by STEMM (where the second M stands for medicine) PhD holders, let us look into some basic details of Alzheimer's disease. According to an early theory, it is caused by misfolded proteins that clump together, killing brain cells, and causing memory loss, and an impaired cognition. A protein, the beta-amyloid, was initially thought to be responsible for this misfolding, but more recently another protein, the tau protein, has been suggested to be the culprit. Unfortunately, most drugs designed to block the misfolding of proteins in the brain have failed.

There are other theories. The neuroinflammation theory states that Alzheimer's is caused by excessive release of toxic inflammatory chemicals from microglia, the immune cells in the brain. Another theory attributes Alzheimer's to synapses, the junctions between brain cells. Yet another theory suggests this to be a disease of the mitochondria, an energy producing structure in brain cells. In every neuron, a fleet of motor proteins ensures the safe passage of cargo from the cell's (neuron's) main body down its axon, or nerve fiber, and back again. That cellular shuttle system supplies materials to the ends of the axon, where electrical signals jump from one neuron to another. Inside neurons, the motor proteins travel along tracks called microtubules. For some long neurons in the spinal cord, the journey from end-to-end could take two weeks. How do these proteins load and lug their precious cargo? They can carry freight many times their size—large organelles such as mitochondria, or proteins contained in bubbles called vesicles. One class of motor protein called kinesins have their two 'legs' plod along the microtubules in eight-nanometer steps, shuttling newly made components to the ends of the axon.

Let us discuss a few other broad health-care problems having potential to generate new opportunities for jobs for STEMM PhDs:

(a) Modern health-care is moving towards implantable closed-loop bioelectronic systems which can monitor different physiological functions continuously and provide quick, real-time, personalized therapeutic solutions. Closed-loop systems are built up from three major components, namely, sensors, signal processors, and actuators. An effective system requires a distributed network of sensors and actuators.

The rise of fiber electronics and AI enables that health-monitoring agents derived on multifunctional fibers should direct future medical and health management to intelligent health. Fabric computing constructs non-chip human-centric sensing, with sensory data from multifunctional fibers distilled by intelligent fabric agents. This will have cognitive applications enabled by integrating fabric computing with AI. One can also do energy harvesting via fabric devices. Application scenarios include health protection, behavior analysis, sleep monitoring, and mental health, by focusing on the fundamental changes that can be brought by the fabric computing framework [11].

(b) Technologies like continuous glucose monitoring are changing the way that a growing number of people manage their diseases. AI will have its biggest impact in biology and health, because biology is so complicated. Only with AI and machine learning can scientists tackle the massive data quandaries that biology poses. To discover the language of life, take a large number of experiments and then use AI to look for patterns that are not apparent to you and me, but make sense to a computer.

(c) Implementation of real-time feedback control strategy, however, is complicated by the complexity of human physiology, which makes it essential to measure and actuate several types of physiological signals at multiple locations throughout the body. This highlights the need to develop a distributed network of bioelectronic sensors and actuators. One generally measures glucose, pH, ECG, EEG signals with appropriate electrical, mechanical, or biochemical actuators to regulate critical activities like neural, functioning of organs, and production of biomolecules. Creation of networks of sensors and actuators that can interface with tissues and organs, presents several challenges such as biocompatibility and specificity to be overcome which needs research inputs [12].

(d) A good idea is to start with early-career researchers (ECRs) working in the STEMM disciplines in universities and research institutes, because ECRs usually have a love of science, which motivates them. It can be PhD researchers or those who have completed their PhD 2–4 years earlier. ECRs often have a challenge regarding career stability, but supervisors do not often discuss career aspirations with ECRs. Perhaps they discuss skill development more.

(e) To avoid public health failures, science needs to initiate implementation research. Implementation scientists move beyond medication and device development and study how to facilitate their use by clinics, front-line health-care providers, patients, communities, and policy-makers. Implementation science would take up proven interventions rapidly. Implementation research would require cross-disciplinary collaborations among scientists who understand communication, marketing, anthropology, economics, and social psychology—disciplines that have not historically interacted with one another [13].

(f) Application of artificial neural networks (AI), are enabling researchers to unlock the complex signaling processes in the brain that handles the processing of language. Brain activity data is collected from study participants who read sentences while undergoing fMRI. These scans showed activity in the brain spanning across a network of different regions—anterior and posterior temporal lobes, inferior parietal cortex, and inferior frontal cortex. Using an AI computational model, researchers can predict patterns of fMRI activity reflecting the encoding of sentence meaning across those brain regions. Such studies help to comprehend how people read sentences, and will contribute to analyzing the brain activity of patients with neurodegenerative diseases when they speak sentences [14].

5.8.2 Agricultural sector

The agricultural sector controls a large percentage of the GDP of most countries, and it offers new career opportunities for PhD scholars, from researching crop genetics, soil science, agronomy, and agricultural economics. Whether it is the production of crops, animal husbandry, horticulture, fisheries, or agro-processing, they all offer a range of career options to suit different educational backgrounds. Options include organic production, beekeeping, dairy production, mushroom production, and polyhouse vegetable cultivation.

Realizing the importance of food and agriculture, graduates and postgraduates in diverse disciplines like management, engineering, finance, and even software, are entering into natural organic farming. There is scope for research jobs in traditional sectors like fertilizers, pesticides, and the development of agri-implements. In addition, new vistas are opening up which would serve the agriculture sector, such as robotics, drones, analytics, sensors, and GPS-driven data acquisition in crop management. All this brings hope for techno-entrepreneurships, catering eventually to sustainability in agri-production.

Computational agroecology combines expertise in technology and farming to develop diverse agricultural landscapes based on natural ecosystems. It is an approach that utilizes computational tools to design diverse, optimal farming ecosystems for sustainable agriculture, and this includes crop selection, soil type, weather conditions, irrigation, and pest control [15]. This topic will assume increasing importance in the coming years due to issues which affect agriculture, such as climate change, depletion of natural resources, soil degradation, and fossil fuel-dependence in farming. Implementation of computational agroecology needs to make available the tools to farmers for planning sustainable farming.

Use of computer technology allows farmers to explore thousands of options to optimize food production without the use of fossil-fuel based pesticides, after varying all the important parameters like soil type, weather conditions, irrigation, fertilization, and pest control.

Globally, 1.6 billion tons of food are wasted every year due to poor storage conditions. Even if 30% of this could be saved by refrigeration, it could feed an additional 950 million people annually, so there is scope for techno-entrepreneurship in this vital sector—food for the hungry mankind.

5.8.3 Chips

STEM PhDs seek jobs in technology. There are going to be jobs for PhDs in the companies which manufacture semiconductor chips. To help reduce vulnerabilities in the crucial supply chain, the US government approved the CHIPS Act, which meant $52 billion in subsidies to companies and research institutions as well as $24 billion or more in tax credits—one of the biggest infusions into a single industry in decades [16].

AI can help generate very impressive images (artwork) in just a few seconds by using software models developed by Google, OpenAI, etc and just a few prompts.

5.8.4 Biology for chemical engineers and sustainability

Each year, factories produce tens of millions of tons of polyester, a non-biodegradable, oil-based plastic material that after use goes to landfills or incinerators. Chemical engineers are working on developing microbes to produce enzymes that can break down polyester into a white powder that can be reconstituted without the hydrocarbons needed to make new polyester. The convergence of biology and sustainability is an under-explored research area.

5.8.5 Dementia villages

Currently, under a new protocol of treatment of dementia patients, facilities are being built which aim at integrating them with the communities around them, thus blurring the lines between home and hospital. For that they try to create an exclusive 'dementia village' [17].

Today there are more than 55 million dementia patients in the world, and WHO expects the number to reach 78 million by 2030. So, this would have scope to employ trained staff—nurses, doctors, psychologists, physiotherapists, and social coaches at such 'dementia villages' or senior 'micro-towns', being located around the globe.

The goal is to emancipate people living with dementia, provide them with the quality of life they deserve and include them in society, essentially marking an evolution from traditional nursing homes, with the difference that the village has a supermarket, a square, a restaurant, etc, to make things as natural as possible. Therefore, the residents can move about freely and mix with fellow patients, getting humanized care that feels like home.

5.8.6 Gene-editing

CRISPR (clustered regularly interspaced short palindromic repeats) is a family of DNA sequences in bacteria that contains snippets of DNA from viruses that have attacked the bacterium. These sequences play a key role in a bacterial defense system, and form the basis of a genome editing technology which allows permanent modification of genes within organisms.

With its huge application value in genome engineering, gene knockdown/activation, disease models, biomedicine, and more, CRISPR technology has the potential to support innovative start-ups. Research in life sciences involves heavy commitments of time in designing and execution of experiments which can be very expensive for large pharma/biotech companies. Start-ups can fill that role.

5.8.7 Biotech with AI

There is a growing trend of young researchers leaving academia for industry. Biomedical postdocs definitely have scope to work for a biotech company, or even have their own start-up in this field, contributing to developing gene-editing therapies for single-gene disorders such as achromatopsia, the type of color blindness which is caused by a single mutation [18], and cancer, or therapies for cardiovascular diseases. Even at pharma companies, biomedical postdocs could find good jobs, which have good salaries, too. Being in such industries is also helpful to transform their conceptual ideas to early-stage clinical tests, an important step in pharma.

With the rapid rise in the application of AI, and given the promising future of bioinformatics and computational biology, the biotech industry is all set for a technological revolution. Look for interdisciplinary programs combining biology and other informatic engineering subjects.

Large pharma/medical companies often depend on small biotech companies to provide good quality R&D results at costs lower than what they would have done themselves with rising overhead costs. It is rewarding for small biotech companies to associate with large pharma companies, health-care, and medtech services to provide IT back-up.

The biotech industry is slated to rely more and more on software, wearable devices, and big data analysis, etc, where IT start-ups can be useful. But the above doesn't have to be at the cost of sacrificing the critical mass of basic science that sustains translational research. That would be harmful for industry, because translational research thrives when academic research with commercial prospects gets transferred to a commercial company.

5.9 New opportunities in STEM jobs by solving existential challenges

The tech sector has been hit by a significant wave of lay-offs. We need to create back-up careers to provide an alternative plan to young academics if their career doesn't work out—a kind of fall-back in a more stable industry, i.e. if their plan A doesn't work out. But then, the young job-seekers are also worried about the cost of re-training for back-up careers, if needed.

On the other hand, digital learning and development tools coupled with online learning can offer a convenient way for people to work on their individual growth at their own pace. In addition to gaining valuable in-demand skills, online learning tools allow workers to try their hand at new job-related tasks, to ensure they are making the right decision before making the leap into a plan B.

The world is challenged by many existential threats, whether old ones or more recent which may amount to a catastrophic situation for humanity. In what follows, we discuss new job opportunities for STEM PhDs, generated by attempts to solve the existential challenges to humanity.

5.9.1 Climate

Economic growth and prosperity have accelerated the use of fossil fuels, resulting in ever-increasing carbon emissions, bringing the threat of climate change closer and closer. Average global temperatures today are more than 1.1 °C (2 degrees Fahrenheit) higher than in the pre-industrial era. A rise of 3 °C by the end of the century looks a possibility. The signs of climate change and the attendant unpredictable and unusual weather events have clearly shown that climate tipping points are much closer than we thought.

Heavy rains and flooding have become regular features across the world, with repeated occurrences. Beijing saw the heaviest rain in the last 140 years on 2 August 2023. Climate warming is driving heatwaves and dangerous wildfires—as extraordinary changes in recent history. Global warming will increase the chance of severe wildfires like those burning across Canada and heatwaves like the one smothering Puerto Rico. Vietnam broke a heat record in May 2023, with temperatures soaring past 44 °C.

Climate anomalies are emerging around the globe and 2023 has seen weather anomalies, some catastrophic in nature [19].

Both in Europe and the US, heatwaves may become hotter by up to 10 °C and some of these heatwaves will last longer and longer every year as the years go by; at least 10 000 people a year die in the US from extreme heat. This year could top this figure in reality. Heatwaves in Europe caused tens of thousands of deaths in 2022.

Extremes of global temperatures being felt currently are consequences of emissions of greenhouse gases, attributed to human activities [20].

2023 has been one of the years with the most heat-related excess deaths on record in recent memory, and July 2023 has experienced the hottest day, and hottest week, measured by modern instrumentation. The heatwaves roiling the northern hemisphere show that the era of global boiling has arrived [21].

There has been a pattern of above-average surface temperatures across global oceans. Much of the excess heat released in the atmosphere is absorbed by the oceans, which is why ocean temperatures have been rising steadily for the past several decades.

Fears about loss of economic stability, lethal diseases, and natural disasters like floods and fires, and the consequent displacement will make humans miserable. These insecurities will be felt by all humans, only that the rich might be able to defend themselves a little while longer than the poor.

Considering the above scenario of climatic emergencies, climate tech start-ups appear to be booming with the potential to bring climate-friendly technologies to the market.

In what follows, we discuss some key aspects of extreme climate after-effects and opportunities for STEM PhDs to contribute to the betterment of humanity.

5.9.1.1 Wildfires

Smoke from wildfires raging in eastern Canada has been filling lungs and turning skies orange across the northeastern United States, most dramatically in New York City and the surrounding area. The warmth and lack of rain has left soils and forests as dry as tinder, so when a fire ignites it can grow and spread quickly—including in places where large, destructive fires are usually rare.

For days together, New York City was bathed in a pall of rust-colored smoke from the Canadian wildfires, an indicator of the worst air quality, in the world. The heavy plume of wildfire smoke from Quebec and some other regions of Canada kept drifting towards the eastern coast of US.

Similar to air pollution, wildfire smoke has nearly 200 different toxins. But a wildfire, takes along with it the smoke arising from the burning of paints and thinners, detergents, shampoo, upholstery, etc, all of which add toxins that affect human health.

While talking of air pollution, one normally talks of $PM_{2.5}$, particulate matter with a diameter of 2.5 µm or smaller. Too small to be seen with the naked eye, these ultra-fine particles go into your lungs. What we smell is the particulate matter, but mostly we smell the volatile organic compounds in the smoke, which are unsafe. VOCs are toxic and go into our lungs, too.

5.9.1.2 Droughts
Residents of western countries use large amount of water per day through showers, toilets, dishwashers, washing machines, and garden hoses. But summers across the world are going to become dryer and availability of water will decrease by 10%–15%.

Many rivers no longer reach the sea. Often artificially straightened and dammed, water is sucked out and channeled off to supply farms, industries, and households. Groundwater is depleting fast. Freshwater is getting increasingly polluted with sewage and fertilizers. Drought is on the verge of becoming the next pandemic, and there is no vaccine to cure it.

5.9.1.3 Heatwaves
Heatwaves are closely linked to droughts. Generally, a large amount of energy from the Sun goes into drying out moisture in the landscape. But as the amount of moisture available for evaporation declines during a drought, more energy is available for heating the air and the temperature rises. During times of heatwaves and droughts, wildfires will ignite more readily, burn more intensely and spread faster.

Heatwaves have been sweeping across eastern regions of China, some other parts of Asia, and even Puerto Rico. Limiting the global warming to a maximum of 2 °C as set by the Paris agreement seems unlikely, and researchers now envisage a potential temperature rise of 2.5 °C–3 °C.

5.9.1.4 Sea ice melting
Sea ice decline is expected to have disastrous consequences for the planet—including the rise of sea levels and disruption of weather patterns. As a consequence of the greenhouse gas emissions the Arctic will lose ice more and more resulting in global warming, increasingly.

North Atlantic Ocean has reached record-high surface temperatures. Sea surface temperatures in the North Atlantic have been higher than any on record.

5.9.1.5 Deep sea mining
The ocean is just a gigantic waste bin to human civilization. How much rubbish, junk, and toxins are dumped in the ocean every single second? With increased warming and slowing of ocean currents, the geochemistry of the ocean is changing rapidly, reducing food stock and nutrients throughout the food chain, and leading to more anoxic zones.

Oceans constitute the main driver of our weather. Now we're adding the threat of mining the sea and ocean beds, on top of the damage we've done with deep-sea oil drilling in the last 40 years.

A new doctoral program has been announced by the Stanford Doerr School of Sustainability. It plans to undertake an interdisciplinary approach to global marine challenges which would comprise of Earth System Science, Civil and Environmental Engineering, Biology and Anthropology. This PhD program would train the next generation of scientists, decision-makers, and policy-makers [22], with the multiple types of skills that are required to address today's formidable ocean sustainability challenges. PhD students will have access to expert researchers and lab spaces on the main Stanford campus and the Hopkins Marine Station in Pacific Grove, California.

5.9.1.6 Air turbulence
Clear-air turbulence (CAT) is hazardous to aircraft and is projected to intensify in response to future climate change. Warmer air will trigger alterations to air currents in the upper atmosphere, known as the jet stream, that will increase turbulence on flights. For airlines, this could mean more wear and tear on planes and higher fuel costs as pilots divert flights to avoid turbulence-prone areas. Passengers and crew may have to spend more time strapped in during flights to reduce the risk of injury.

5.9.2 Looming AI

AI and machine learning (ML), are new technologies that can make a device make its own decisions, not necessarily sticking to the set of instructions given to it at the time of training. In this way, it takes something away from human beings. These new technologies will have autonomy, and so they will ultimately be able to learn and act on their own. They will dominate humans and might ultimately displace us (perhaps) from being the most influential species on the planet.

5.9.3 Smartphone-distracted driving

Distraction causes a huge number of crashes of vehicles while being driven. Screen addiction has added a new dimension to the problem. Therefore, why not let a smartphone become a tool to combat device-based distraction. When air-bags can open instantly, why can't smartphones have safety feature apps to alert a driver using it and driving simultaneously of an impending disaster like hitting a pedestrian or a stationary vehicle, etc. There already exist 'do not disturb' features that can block incoming calls and notifications while the user is driving. Other apps have the potential to reduce other types of distracted driving, encourage safer speeds, and provide basic crash avoidance capabilities. For that the smartphone must monitor the driver continuously and incentivize safe driving by rewarding drivers with lower premiums.

The smartphone camera should monitor the driver's gaze or head direction and alert them when their attention drifts from the road ahead for too long. The marketing companies should provide forward collision warning (FCW).

5.10 Opportunities for new jobs in the electrical vehicles sector

Tesla was a champion of wind and solar power, suggesting them to be superior to coal, and he dreamt of machines powered just by gravity and frictionless engines. Now, an EV automotive company bears his name [23].

By 2030, the world plans to move to making at least 50% of all new car sales be electric, and phasing out ICE-based gas-powered cars totally by 2035. Endangering of pedestrian safety due to large EVs coming onto the market, calls for research into how vehicles are designed—their size, or composition, their weight, their proportions, even their shape. New jobs will emerge here. Also, large EVs mean large batteries which implies more extraction of minerals like lithium and cobalt, while charging them would require additional electricity. Their extra weight also generates more particulate emissions from brake pads and tires.

New jobs are bound to arise in the electric vehicle sector, or the battery-operated automobiles. Norway, which began promoting electric vehicles in 1990s, has already seen sharp falls in levels of nitrogen oxides, by-products of burning gasoline and diesel that cause smog, asthma and other ailments, as electric vehicle ownership has risen. Noise has reduced so that heavy electric equipment like earth excavators can work quietly even in the vicinity of schools or hospitals. Thus, electric vehicles are part of a broader plan by Oslo to reduce its carbon dioxide emissions to almost zero by 2030.

5.10.1 Jobs being created by EVs in other sectors

Electric vehicles are creating jobs in other industries. (i) Fast charging stations in the public areas and developing a charging infrastructure that works flawlessly with the different software used by dozens of kinds of vehicles. In far denser Urban Areas like NYC, finding an open charger may prove nearly impossible. (ii) Servicing of electric vehicle batteries. (iii) Servicing of electric vehicles which, although they don't need oil changes and require less maintenance than gasoline cars, still break down. (iv) Tackling microscopic particles resulting from the abrasions of tires and asphalt, levels of which have been rising due to increasing use of heavier electric vehicles than internal combustion engine cars. Heavier means they cause more abrasion of tires.

Artisanal mining of cobalt, tin, gold, and tungsten is often associated with pollution, hazardous working conditions, and child labor. More than half of the world's cobalt production comes from mines in Congo (DRC), of which 20% is extracted by artisanal miners under hazardous working conditions. Cobalt is toxic, and people living near such mines have been found to have high levels of the metal in their urine and blood.

China refines an estimated 58% of the world's lithium and 65% of the world's cobalt, much of which is mined in the Democratic Republic of the Congo by Chinese-owned companies.

The story is the same when we consider computer chips and batteries needed to assemble smartphones and computers, for which refineries and smelters in various countries supply minerals like cobalt, lithium, tin, tantalum, tungsten, and gold.

Rather than going into landfills, recycling of discarded batteries would help reclaim the key metals lithium, nickel, and cobalt to go back to lithium-ion battery manufacturers, helping reduce the cost, and protect the environment by saving the efforts needed for mining them afresh.

5.10.2 Jobs from tackling the devil in the EVs

A sure-shot job opportunity would be in a battery recycling center. Used battery packs are dismantled and shredded by a machine to separate plastic, aluminum, and copper from a black mass that contains crucial ingredients such as lithium, nickel, cobalt, manganese, and graphite. That is how the recycling of batteries will greatly reduce the need for mining.

   A. depending on how aggressively they charge their EV's, the battery wears out in 6–8 years and the quoted cost to replace it then will be greater than the actual value of the car. Using super chargers (SC) to charge an EV battery makes it to wear out sooner. This is because super chargers heat up the battery significantly and the customer is advised to wait until the battery cools down. But people are in a hurry and do not usually wait for the cool down, thus contributing to shorter battery life.
   B. Anything that uses electricity in an EV reduces the range of the car. Using an EV in cold weather with the heater running cuts available distance by 40% or more. Towing something reduces the range, too. Manufacturers warn to recharge when the battery hits 30%. That means a 300 mile range is actually a useable 210 miles. And that's for a brand new vehicle. With 30% battery wear, you are looking at a usable range of perhaps 140 miles between full charges with that number on an ever-declining downward slope as the battery accumulates more wear.

5.10.3 Scope for R&D in the EV sector

There is only one proven viable, scalable, and technologically ripe scheme for decarbonizing personal road transport. That is electrification.

If the electric vehicle transition is further delayed, there are likely to be cascading effects elsewhere that will ultimately put a brake on global decarbonization. The demand for personal powered mobility is increasing in low- and middle-income countries. In Asia alone, cars are projected to account for more than 10% more trips taken in 2050, vis-à-vis in 2015. On the basis of current trends, there will be three billion cars and vans on the road globally in 2050, up from one billion now—another reason to accelerate the transition to electric vehicles worldwide.

A problem lies in the phrase 'carbon-neutral fuels'. These fuels rely either on inputs such as 'green' hydrogen, which is made by splitting water using renewable electricity, or on feedstocks such as biomass. The technologies used to make these fuels are inefficient, expensive, and untested at scale. Moreover, claims of climate neutrality—based on the idea that the CO_2 emitted by their combustion was absorbed relatively recently from the biosphere, or that CO_2 produced during their manufacture was prevented from entering the atmosphere—are questionable [24].

The research community must be equally clear in underlining why this needs to be addressed.

5.11 The new scenario under the emergence of AI and ChatGPT

On November 30, 2022, OpenAI launched the latest version of ChatGPT, the largest and most powerful AI chatbot to date. Millions tested its ability to do the mundane things like writing emails, coding software, and scheduling meetings. The reality is that ChatGPT is a game changer which has shown the potential of AI to change the way we live and work.

ChatGPT, a text-generating AI built by OpenAI, can crank out manuscripts, original poetry, and working examples of code. AI can help generate very impressive images (artwork) in just a few seconds by using software models developed by Google, OpenAI, etc, and just a few prompts.

AI will continue to evolve, eventually taking over many jobs that are currently performed by people, it will also create many work opportunities that don't yet exist. Companies need to learn to do things differently from the way they've been done over decades, and they must transform the skill base. The next section discusses how Unilever attempted to achieve this by undergoing end-to-end change from the manufacturers right down to the consumers or the products.

Let us dig deeper into recent forays of digital technologies—how they influenced ways of life of humans and even destroyed some existing humanistic experience, so that we can learn to keep a sharp focus on how to handle the emerging digital world and the jobs scenario for STEM PhDs.

Microsoft made electronic records mandatory, and now doctors can't talk to each other and the patients suffer. Amazon destroyed the in-person shopping experience and lead to dead malls all over the world. Google is fun to look up everything instantly, but I used to do the same with Encyclopaedia Britannica, and it was equal fun in those days. And that didn't throw 'personalized' ads at us. Online buying increases pollution with shipping, monopolizes industry, mostly for the lazy. 'Cloud Storage' is a server, and nothing new about it.

There has been a boom in 'generative AI'—a term for the new type of artificial intelligence apps, trained on vast amounts of data, that can create new media objects. How far generative AI will influence our lives, cannot be said, but some people argue that these apps will destroy millions of jobs, while others argue that they'll be a boon to human creativity. To put it in a nutshell, the AI tools have arrived, it is for us to decide how to use them.

5.12 A case study of Unilever

The speed of delivery of our packages has till now been to go to the store and pick this up and the store can replenish itself over a week-long horizon. But now we just press the button in the app and expect it to be handed to us in the next five minutes. So, an organization needs to function differently, and attend to the customization of the personalized products that consumers will require. Then under Managing the

Future of Work, we have tech and demographics coming together in the gig workplace.

A massive change is that direct-to-consumer is a channel now, just as we order stuff off Amazon directly. Getting to consumers can be done by any little start-up now by throwing in some ads on social media, speak to a few influencers and start sending their products out.

Unilever is in a fast-moving consumer goods (FMCG) business, serving 3.4 billion people every day through its 400 brands/products across 190 countries. It has 149 000 people employed directly by Unilever, and an extended workforce of 3–5 million people employed by others but working for Unilever.

Large manufacturing companies have a large number of full-time employees working under a lot of security, but no flexibility in how and where they work. On the other hand, there are gig workers, freelancers with a lot of flexibility in working conditions, but hardly any security about guaranteed income. To this end, Unilever launched its Future of Work initiative in 2016 to accelerate the speed of change in its workforce for a digitalized and highly automated era. As of July 2023, the program showed some success but still faces challenges in implementation [25]. Change is happening all the time as all Unilever's factories are rapidly automating their office processes.

Unilever has been exploring whether they can create an alternative by bringing in more flexibility to their workers in choosing how and when they work. So, they tried U-Work, where employees have no job title. They work on gigs but are not gig workers but they are still fully Unilever employees. They get a guaranteed retainer and a package of pension and health-care benefits. They only need to work on projects, but are free to do their own business on the side, and have the flexibility to look after their kids or aging parents.

So, Unilever has to get into a new contract with such workers wherein Unilever sets out to define what role an employee would have in the future, and what skills he would need to get to that future point. Of course, Unilever would give them the platform to acquire those skills. The employee would have to put in the time and investment to acquire those skills when that time comes. The only catch is that Unilever would not guarantee lifelong employment.

Unilever don't think jobs are really going away because jobs are a collection of tasks. Certain very manual tasks will be taken over by machines and require less human input as the technology gets more advanced. AI, ML, etc are able to take those off.

PhD students and degree holders should offer to innovate and make alterations in the business model for firms which have a forward-thinking mindset and can invest a bit to beat a sudden disruption, but do not have to do so themselves or think why bother if all is going right and their business is making money.

Digital fluency and also new skills are important for the future, but we also need judgment and innovation. Developing proprietary data handling will be done by humans who will be complementary.

There is no point in stating that you had better re-skill or the robots are going to take your job away. Instead, a company has to motivate the employees to learn new skills. Unilever is talking about ensuring that 80%–100% of the workforce of 149 000 employed directly by Unilever can be transitioned.

5.13 Job security

The tech industry is fragile and offers zero job security. Whenever a recession hits it affects the tech industry. Job security is a myth. Always be ready to leave. Your main focus should be on keeping your skills up-to-date, your network warm and your eyes and ears open. Help friends who are looking, because you might be next. Be proud of what you've accomplished so far, but continue to invest in yourself. If you like what you are doing, and have a direct pipeline to profits from your labor, you can weather any storm.

Stability in jobs is ephemeral. Just when you get to your best place in your career, the rules change. New technology has been the biggest game changer. Be flexible; life is full of surprises, some of them nasty, but enough can be pleasant for those who are prepared to take a leap. Plan for the best and prepare for the worst, for your boss is not your friend and no corporation will ever have your best interest at heart.

Take a job where you will learn the most, not the job that pays the most. STEM PhDs should be choosy in taking up their first job. It will influence your second, and third, and so on. Be reliable. Be a team player. Listen and learn. Say yes to assignments. Don't expect instant promotions. Have faith that you have several skills. Learn to deal with stress. Go for excellence, humility, positivity, team orientation. And when you wish to criticize any point or statements, you should also have a suggestion on that, or a solution for it.

The job market goes up and down. Better times will come; be strategic. Worse times will come; be strategic. Recognize that you are always replaceable at work.

Education should give us not only skills but wisdom, the wisdom to recognize that we cannot predict the future, difficulties will arise and preparing ourselves to maintain balance and inner strength in the face of life's uncertainties is vital. So, for stability, focus on your inner resilience.

The college degree is just a baby step. Performing well at work over many years is the key to getting what you want. There is no shortcut.

Make sure your network includes a majority of people younger than you. If your network is all older, they will retire while you still need connections.

Prepare to commute to the office, factory, warehouse, etc, to interact with real colleagues and real infrastructure. The days of sitting home and 'working' at a computer screen remotely are drawing to a close for most.

Take failure in your stride. PhD students and postdocs should learn to weather dispiriting setbacks. The camaraderie of commiserating together over frustrations and unsuccessful experiments helps all lab members tremendously. Lab members helping each other leads to successful experiments and publications, faster. Let postdocs break out into small groups so they can talk about their problems peer to peer.

Finally, it is worth stating that in the war of the currents, Thomas Edison's DC power was pitted against Nikola Tesla's patents for alternating current (AC). One of

Tesla's fondest dreams was the transmission of power from station to station without the employment of any connecting wire! Tesla once thought why has one to fuss with laying telegraph cables (or power lines) if the earth itself could deliver electrical signals of varying strengths to a specific addressee? He was obsessed with the idea of effecting communication to any distance through the earth or environing medium, and extraterrestrial recipients. Let us keep that in mind when looking at the possibilities of tech-innovations yet to come!

References

[1] Muratbuffalo blog 2020 How innovation works and why it flourishes in freedom (2020, by Matt Ridley) https://muratbuffalo.blogspot.com/2020/11/how-innovation-works-and-why-it.html
[2] Owens B 2022 Canada announces new innovation agency *Nature* **April 2022** https://doi.org/10.1038/d41586-022-01190-4
[3] The Editorial Board 2022 Save America's patent system New York Times https://nytimes.com/2022/04/16/opinion/patents-reform-drug-prices.html
[4] The Chronicle of Higher Education Content sponsored by VCU 2020 https://sponsored.chronicle.com/unleashing-innovation/index.html
[5] Gorvett Z 2021 The ancient fabric that no one knows how to make *BBC* https://bbc.com/future/article/20210316-the-legendary-fabric-that-no-one-knows-how-to-make
[6] Guo C F and Ding L 2021 Integration of soft electronics and biotissues *Innovation* **2** 100074 https://cell.com/the-innovation/fulltext/S2666–6758(20)30077-1
[7] Ekblom J and Paulsson L 2023 Sweden's $100 billion green tech boom *Bloomberg* https://bloomberg.com/news/articles/2023-02-15/subarctic-sweden-is-at-the-forefront-of-a-100-billion-green-tech-boom
[8] Scott S 2023 The Handoff *Stanford Magazine* https://stanfordmag.org/contents/the-handoff
[9] Rajkumar K *et al* 2022 A causal test of the strength of weak ties *Science* **377** 1304–10
[10] Guo H *et al* 2022 Further promotion of sustainable development goals using science, technology, and innovation *Innovation* **3** 100325
[11] Chen M *et al* 2022 Fabric computing: concepts, opportunities, and challenges *Innovation* **3** 100340
[12] Bhave G *et al* 2021 Distributed sensor and actuator networks for closed-loop bioelectronic medicine *Mater. Today* **46** 125–35
[13] Proctor E K and Geng E 2021 A new lane for science *Science* **374** 659
[14] Anderson A J *et al* 2021 Deep artificial neural networks reveal a distributed cortical network encoding propositional sentence-level meaning *J. Neurosci.* **41** 4100–19
[15] Runck B *et al* 2023 State spaces for agriculture: a meta-systematic design automation framework *PNAS Nexus* **2** pgad084
[16] Swanson A and Clark D 2023 Chip makers turn cutthroat in fight for share of federal money *New York Times* https://nytimes.com/2023/02/23/business/economy/chip-makers-fight-federal-money.html
[17] Plockova J 2023 As cases soar, 'dementia villages' look like the future of home care *New York Times* https://nytimes.com/2023/07/03/realestate/dementia-villages-senior-living.html
[18] McKyton A *et al* 2023 Seeing color following gene augmentation therapy in achromatopsia *Curr. Biol.* **33** 3489–94

[19] Dennis B and Dance S 2023 It's not just hot. Climate anomalies are emerging around the globe *Washington Post* https://washingtonpost.com/climate-environment/2023/07/31/july-hottest-month-extreme-weather-future/
[20] Forster P M *et al* 2023 Indicators of global climate change 2022: annual update of large-scale indicators of the state of the climate system and human influence *Earth Syst. Sci. Data* **15** 2295–327
[21] Milman O 2023 'Silent killer': experts warn of record US deaths from extreme heat *Guardian* https://theguardian.com/us-news/2023/aug/01/heat-related-deaths-us-temperatures-heatwave
[22] Clancy M *et al* 2023 To speed scientific progress, understand how science policy works *Nature* **620** 724–6
[23] Dukes H 2022 Earthen messages: Nikola Tesla in his laboratory (ca. 1899) *The Public Domain Review* https://publicdomainreview.org/collection/nikola-tesla-in-his-laboratory/
[24] Gupta A 2023 European backsliding on electric vehicles is bad news for the climate *EQ Mag.* https://eqmagpro.com/european-backsliding-on-electric-vehicles-is-bad-news-for-the-climate-eq-mag/
[25] HBS 2023 *How Unilever Is Preparing for the Future of Work* (Harvard Business School, Working Knowledge) https://hbswk.hbs.edu/item/cold-call-how-unilever-is-preparing-for-the-future-of-work

Chapter 6

Models adopted to upend education for supporting growth of graduate jobs

In this chapter we discuss how employability of STEM graduates and PhDs can and is being enhanced through skill-development programs, and multidisciplinary education efforts.

We describe, with examples of how some countries and institutions are specifically attending to this, and how new opportunities are being created for jobs for the disabled, as well.

6.1 Introduction

There's a huge gap between being qualified and being employable—and that gap is what skilling programs try to fill. Skills become obsolete when the pace of change is fast. If you're not learning anything new in, say 10 years, you are virtually obsolete and no longer employable. Hence, the game is all about rapidly accessing the new knowledge and internalizing its learning. Learning, in fact, continues all through life, because it is not a finite activity, which when once over, you are done.

A wide range of disciplines, from humanities, social sciences and languages to natural and physical sciences are truly interconnected, a fact that can be appreciated by a student at a university teaching liberal arts.

Study and qualifications have been the path to the job of a student's dreams. But, the link between attainment of higher education qualifications and the movement into certain professions is no longer happening so easily. In most disciplines, the number of PhDs awarded each year outstrips the number of available tenure-track positions, bringing about an oversaturated job market where finding employment in academe is not easy.

In November 2022, the world's attention was brought to the launch of a chatbot: ChatGPT, created by San Francisco-based OpenAI, and within a week it had already reached over a million users. ChatGPT is just a tool, and a tool can be used in both good or bad ways. It is an example of how the landscape is changing, not only in the way we interact and work, but also in the way we educate ourselves, which in turn is being shaped by higher education.

Unexpected crises such as the pandemic and the war in Ukraine create unemployment shocks. The market's slowdown, post-Covid, has made companies to put off projects and shift priorities. They're more cautious with their expenses. Challenges in fields such as agriculture, food, and the environment need technology-based solutions. An ecosystem of entrepreneurs, investors, and researchers, alone can provide those solutions.

6.2 Jobs crisis post-PhD, and how to fix it

6.2.1 Multi-disciplinary education

Why, what, how, and where to receive the higher education is more and more being linked to the job aspirations of students, and not much to the desire to learn. Employability and job security of a candidate is now linked to inculcating a multidisciplinary knowledge-base among students of all disciplines, whether it be arts, science, law, engineering, medicine, or commerce.

With digital technologies unfolding exponentially, it is imperative that teachers/academics re-orient themselves quickly because the days of a professor walking to the podium of a classroom filled with a hushed audience of students are disappearing fast. The internet brings so much awareness to all, including the students, that teaching and learning is mutual and is becoming a preparation for the fast-changing job scenario. Only the fittest with sound interdisciplinary talents will be on top of the situation, and that applies to both the teachers and the taught.

To cater to the fourth industrial revolution, multidisciplinary education is central, simultaneously incorporating vocational subjects and soft skills. Science and engineering students should spend more time on learning arts and humanities, while arts and humanities students should aim to learn more of STEM subjects, at least as much as possible. Advances like big-data analytics, machine learning, and AI demand a workforce skilled in mathematics, computer science, data science as well as in multidisciplinary abilities across the sciences, social sciences, and humanities. A typical example is that of the complex problems in health-care which cannot be addressed successfully by a single discipline, and needs the promotion of interdisciplinary research culture by developing strong collaborations between medicine, health, and case studies worked out by using mathematics, physics, and engineering [1].

6.2.2 Strengthening the skills of writing and drafting

It is often observed that STEM experts, particularly those with a PhD degree, draft their writings with more clarity than others. This art and maturity comes to scientists/engineers after spending several years in their profession, because while writing their research work into a manuscript, it must be stated unequivocally. Otherwise, the manuscript might invite serious queries by referees and reviewers.

For young STEM researchers, however, the drafting of a manuscript of their recent professional work, or writing a cogent funding proposal is a big challenge. An exquisite writing on any subject is that which cannot be further improved upon. This comes from several revisions of a draft until it has the minimum number of words which are essential, and no more alterations can be made to it. They take time to learn to use the appropriate phraseology, and need to avoid being repetitive while writing any arguments. Help taken from the fast digital technology (ChatGPT) in drafting takes the 'beauty' out of the narrative. To solve this issue, there should be a course on 'Experimental Writing for Scholars', taught to them in order to orient them gradually to learn to write a mature narrative, which is exact, graceful, and unequivocal. And that is true for all the subjects whether STEM, arts, humanities, or commerce, because all of them require writing of statements that have clarity and a focus that stands out.

6.2.3 Mathematics is essential for employability

Math courses present 'the most significant barrier' to degree completion in both STEM and non-STEM fields. Without algebra, you can't solve any problems involving an unknown factor. Without trigonometry and geometry, you can't lay out a garden. Without calculus, you cannot know where the outfielder needs to be to catch a flying ball. Any graduating high school student should know the fundamentals of these math skills.

Math needs to be taught a lot better if we want children to like and understand it in order to make successful careers. In some countries, mathematics and arithmetic teaching is separated so that those who want to continue on and enjoy trigonometry can, and others can focus on counting, ratios, fractions, etc, that are more commonly needed in daily life.

Anything beyond the most basic algebra and geometry should be elective. It is only going to be used by those who specialize.

Multiplication tables are a fundamental underpinning for success. Most young students know their 2×, 5×, and 10× tables but beyond that a majority of them virtually hit a wall. And many would immediately reach for their calculators to find the answer! There are benefits of mastering the tables—today and for the rest of their lives. The only resistance to any of this is from the 'educators'. They object to what is called memorization (and they do so by referring to it as 'rote learning'). Every student can use their mastery of the tables in everyday life from supermarket shopping to every other transaction. But, until school boards make it acceptable and common practice in the classroom, memorizing will not happen. And, why not keep the phones and calculators out of the math class?

6.2.4 Financial literacy

It is becoming increasingly important to consider the subject of 'Instructions in Personal Finance' as a requirement at high school graduation. Universities, including Stanford, are now offering personal finance courses. Employers, too, are recognizing that financial anxiety hurts employee productivity and are sponsoring personal finance lessons at the workplace itself.

6.3 Country-wide efforts

In this section, we discuss efforts being made in some countries to boost the employability of STEM graduates and PhDs through both education, and learning of skills.

6.3.1 France

Sylvie Retailleau, the Minister of Higher Education and Research in the French government, a physicist by training, plans to shake up French science by striking a balance in it between research, training, and innovation. She has undertaken to establish closer collaboration amongst experts in the area of education, health, agriculture, and industry. To establish a long-term plan, she has called for the state to invest an extra €26 billion in research by 2030, a good chunk of which would be spent on health innovation and research. Salaries would be increased to attract young talent to careers in STEM. Management of laboratories will be simplified in order that researchers have more time to devote to research. Management tools would be shared among labs' supervisory bodies to avoid duplication of administrative tasks, and to simplify the introduction and follow-up of research programs.

She plans to change the present scenario where many schoolteachers in France have no science background, by making science to be a compulsory subject in the teacher training bachelor's course, from the academic year 2024–25 [2].

In order to make qualifying, certification, and diploma training accessible, which until now was mainly provided face-to-face within universities, FUN (France Université Numérique, the Digital University of France) is offering, in collaboration with the UniCamp, the 19 universities collective, a diverse range of online professional training alongside the traditional MOOC offers, from 2024. The FUN offer implies training allowing the validation of skills in the form of micro-certifications. These modules will allow learners to target specific skills to acquire, thus offering a personalized and modular approach to the acquisition of professional knowledge. Training needs on the job market are deeply influenced by the rapid evolution of technologies, the transition to a digital economy, and changes in the nature of work. Therefore, the training programs must also adapt to meet the specific needs of fast-changing scenarios. The emphasis of FUN would be on continuing training, professional retraining, and lifelong learning which are becoming crucial to enable workers to adapt to the job market [3].

6.3.2 India

There are 993 universities, 39 931 colleges and over 10 000 stand-alone institutions in India where over 37 million students are enrolled. Out of these, almost 80% enroll for undergraduate courses, of which 16.5% are in the sciences. That makes about 500 000 science students. Out of the colleges, about 60% or around 24 000 colleges are in the rural areas.

By establishing a 'National Research Foundation (NRF)' with a budget of about US$6 billion over five years, India plans to increase the spread of a research culture across the nation's hundreds of universities, colleges, institutes, and laboratories. A very small number of roughly 40 000 higher education institutions, run by the states in India conduct research. In excess of 95% of higher education students go to state-funded universities and colleges. On the other hand, about 65% of the funds provided by the Department of Science and Technology's Science and Engineering Research Board (SERB), one of India's major research funding agencies go to the Indian Institutes of Technology, which are owned by the federal government [4]. The NRF's goal is to 'seed, grow, and promote' research across the country's institutions by strengthening ties between academia, industry, and the government.

For decades, students and staff from the first generation of IITs have excelled at US universities and in Silicon Valley companies, something that has been repeatedly acknowledged as 'brand IIT' by business, political, and scientific leaders, including former US president Bill Clinton, as well as Amazon and Microsoft founders Jeff Bezos and Bill Gates. The number of prestigious Indian Institutes of Technology has trebled in the space of the last decade.

Some IIT graduates will no doubt want to follow in the footsteps of their alumni such as Sundar Pichai, chief executive of Alphabet Inc. and its subsidiary Google, Satya Nadella, CEO of Microsoft, and Arvind Krishna, IBM chief executive. But the overwhelming majority are building and working in companies at home. In the early years after the inception of the IITs and almost till the later 1990s, a very large fraction—sometimes as high as 60%–70%—used to go abroad. Now the numbers are down to a few per cent.

The National Education Policy (NEP)—2020 of India calls for technical training, such as that provided by IITs to engineering graduates, to be interwoven with 'opportunities to engage deeply with other disciplines', with the aim of eventually 'enhancing the employability of the youth'. Undergraduate students at IITs do three electives from humanities in a year, from subjects like anthropology, sociology, linguistics, literature, history, politics, philosophy, economics, and public policy.

Very recently, a pioneer in education of engineering technology in India, BITS, Pilani, having its campuses at four places, namely, Pilani, Goa, Hyderabad, and Dubai, has opened a fifth one near Mumbai, to cater exclusively to impart education in management to its engineering graduates.

6.3.3 USA

The CHIPS and Science bill passed by the Biden government is to develop semiconductors which would lead to investments in climate and energy technologies, with an eye on providing an opportunity to build industries and jobs, right within USA.

In 2024, Golden Gate University (SF, CA) has introduced a 3-year program of Doctorate in Emerging Technologies with Concentration in Generative AI [5]. A focus of this course will be on the emerging technologies like Generative AI, encompassing cutting-edge advancements such as Large Language Models, Brain Decoding, etc. This program is designed specifically for business leaders keen to learn the intricacies of Gen. AI and benefit from the ongoing technological revolution. PhDs from this program will not just be aware of the know-how to implement AI but will also be in a position to offer a strategic vision to set up the right AI strategy, align structures, and address ethical considerations at their job and workplace.

6.3.4 Germany

Germany started 2024 with Berlin's streets choked with tractors and farmers blaring horns in furious protest of proposed budget cuts. Then train engineers walked off the job to demand better pay, stranding commuters and carloads of freight and leaving the country angry and gridlocked [6]. German recession can be blamed on the war in Ukraine, which abruptly ended the flow of cheap Russian natural gas to German homes and businesses. Shutdown of German nuclear power plants has also led to high energy prices and greater fossil fuel pollution. It is hoped that increasing adoption of AI in the coming years will eventually create new jobs, boost GDP, productivity, and stock prices for Germany.

6.3.5 China

The American economy has outperformed China post-Covid. China's economy, which only a few years ago seemed headed for world domination, is not doing so well. The one-child policy combined with the traditional preference for boys has created a large cohort of unmarried and unmarriageable men. The female population sees little hope for their future and are not having children. China now also has an aging population, combined with a declining population growth rate.

China has two consumer markets. Domestic and foreign. Its real estate debacle has weakened its domestic market and its foreign policy has shrunken foreign markets. However, China has the industrial capacity to own the electric vehicles market in many countries around the world. Even in factories making electric cars, robots do most of the work. Chinese factories are insanely efficient, as far as human labor goes, robots doing much of the assembly. Young Chinese have limited job prospects

6.3.6 UK

UK universities have pioneering research institutes offering opportunities for doing PhD to students interested in pursuing a career in academia.

The strong industry-academia linkages in UK draw international students to the courses tailored to meet industry-specific needs, such as Computer System Engineering, which has a variety of modules like mechatronics, machine learning, control theory, and system design. These modules have applications in diverse industries, including biomedical, aerospace, finance, etc. Computer Systems Engineering graduates can go for further specializations, including as control and software engineers, systems integration and testers, software developer, and project manager.

However, Inflation is driving up operating costs in UK universities. International students, typically pay about double the home fees in UK, and are the primary source of additional income enabling many universities to make ends meet. For a long time, this money helped fund research, but it is now being diverted into making up the shortfall on domestic undergraduate tuition [7].

6.3.7 Malaysia

An IAEA initiative in Malaysia is geared to create greater inclusivity in STEM education with the support of a regional IAEA technical cooperation (TC) project on Strengthening Nuclear Science and Technology (NST) Education at Secondary and Tertiary Levels [8]. Aimed at using the role of nuclear science in achieving several of the United Nations Sustainable Development Goals (SDGs), particularly in health, food security, energy, and environmental sustainability, the project has a goal of reaching 10 million students in Asia and the Pacific, working towards universal inclusiveness in nuclear science education.

By integrating empowerment and enablement strategies, this TC project acknowledges the unique strengths and potential of special needs students in STEM fields, seeking to unlock their talents and foster their interest in nuclear science. It focuses on the development of learning materials, tailored to the diverse needs of special needs students. The hands-on learning experience offered in the nuclear science and technology program is going to be invaluable.

6.3.8 Vietnam

Vietnam is investing in training and education, with a view to becoming a leader in electric vehicle manufacturing. For that, efforts are being made to keep the pace of STEM education with the development of technological innovations.

6.4 Enhancing employability of STEM graduates and PhDs

6.4.1 By staying on top of technology (example—electronic design engineers)

According to a World Economic Forum (WEF) analysis, over 85 million jobs will be displaced by 2025 due to the rise in automation while 133 million new jobs are expected to open up due to new advanced technologies and digital transformation. In such a scenario, new roles emerging due to automation will require new skill-sets and hence upskilling and reskilling of the existing scientific workforce.

Electronic design engineers, for instance, are frequently called upon to accommodate accelerating change, with no time to lose. Although a majority of these engineers hold a Master's degree or a Bachelor's degree, education levels are improving, with about 13% of them holding the highest level of education, i.e., a PhD degree. For most electronic design engineers, there is no getting around the need to refresh one's knowledge and skills in emerging technologies, such as machine learning and artificial intelligence [9]. To do that, their preferred means of getting up to speed on technology was engineering videos, at least until 2023. While fears of contracting disease (Covid-19) have largely receded, lack of time remains a problem for the knowledge hungry, who are keen to attend shows, conferences, and seminars.

6.4.2 By coaching them young

A majority of children think originally (unlike each other) before they enter school. But, by the time they reach about 25 years of age, only a miniscule number of students think differently, the rest of them think alike! This is reminiscent of the 18th century Industrial Revolution, during which a mass education system was designed to prepare children to work in factories, like robots, and not to think. The children became devoid of freedom to think, and their individuality. It continues. Only industries have changed, and the focus has shifted—now to AI.

It is important to realize that assessment drives learning. Students will learn the way they are assessed. One route is to gradually draw our students away from the traditional rote learning based education, and orient them towards innovation and creativity. Therefore, questions in exams need to be on concepts, critical thinking, analytical thinking, and different ways of looking at avenues of knowledge and not just on speedy reproduction of information.

Science is mainly an experimental field of study. Video demonstrations and virtual labs with simulations need to be integrated along with hands-on experiments for better understanding and visualization of what is happening inside the reaction flask or apparatus!

Doing science requires authentic communication. Therefore, when students attempt to explain phenomena or solve problems, they simultaneously improve their language skills. So much of learning is about storytelling, coming together, and having conversations. While being in a class together the students can thus learn from each other.

Today's students face a job market where employers want a broader skill-set, not just graduates with a degree in a particular field. Schools that focus heavily on academic performance and not enough on careers counseling are doing their students a disservice. Disarm students to engage them, by giving them something NEW enough to steer them away from distractions like the smartphone. They should feel excited about the new methodology of teaching, which therefore, must appeal to them. We need subject teachers, who interact amongst themselves by congregating regularly, spontaneously, to devise methodologies to impart project-based learning, with problem solving, and problem designing by students themselves.

We require a cadre of talented scientists, engineers, and other STEM professionals to advance knowledge, and design new technology of education. Science education needs to stimulate children's intellect and imagination and motivate them to consider science as they work to solve the pressing problems the world faces today and will confront tomorrow: cancer, future outbreaks of disease, tenacious agricultural challenges, climate change, food insecurity, and disparities in health and wellness between racial groups, to name just a few.

Taking a typical example of say banking jobs, all the highly successful people there are very bright. Some of them have a Masters or PhD in another field like music or linguistics, and have then switched to programming to suit the job requirements. High-tech companies, too, have people with a Masters or PhD in Computer Science from say, MIT or Berkeley. In current times, STEM PhD holders need to be aggressive learners who can turn their hand to whatever is required and get the job done without any supervision. Let us not forget that three-fourths of PhD graduates are employed outside universities. It's vital, therefore, that students receive high-quality information about alternative careers.

6.4.3 Through institutional efforts

The world's greatest challenges pose multidimensional issues, whether it be climate change, wiping out hunger, or poverty, they all need collaborations across multiple disciplines. We need to foster innovation and facilitate collaboration by erasing artificial barriers among disciplines. This requires an interdisciplinary approach to developing big ideas, facilitating conversations, and promoting execution beyond the traditional boundaries of education. Faculty, students, and business leader mentors can interact regularly to explore how ideas become reality and problems have solutions. Motivated teams should have the chance to earn seed money and transform their innovative ideas into prototypes or pilot projects.

The Stanford Woods Institute for the Environment is awarding up to $200 000 per project focussed on interdisciplinary research required to solve major environmental challenges, too complex for any one discipline alone to tackle [10]. Three examples of projects being supported by Stanford Woods are described below.

6.4.3.1 To make cities healthier
Connections are to be found between nature in urban areas and human health. An interdisciplinary team integrates a wealth of new data, science and analytics from San Francisco, Singapore, and Guangzhou. Urban planners then devise a policy on how to enhance urban Nature and derive its benefits, and improve health, equity, liveability, and the sustainability of cities. (For experts in biology, epidemiology and public health, and pediatrics.)

6.4.3.2 To study global warming
Sensors are to be developed and strategy to deploy them devised to monitor gas pollution in the Arctic. Using optical spectroscopy with inexpensive, passive, and biodegradable sensors, placed in corner-cube arrays that act as retroreflectors, whose signals would be remotely detected by spectroscopic interrogators placed in stationary base stations, or on movable platforms, like airplanes and drones. (For experts in electrical engineering and applied physics.)

6.4.3.3 Protection of women's health
From heavy metals in the environment which are considered a health threat to women during pregnancies in poor countries, leading to still birth, etc. Working in

Bangladesh, the researchers evaluate concentrations of various metals in drinking water, soil, rice, and turmeric to identify the likely routes of exposure to heavy metals during pregnancy, and identify pathways connecting environmental metals to still birth, and suggest interventions for improving women's health. (For experts in medicine, pediatrics and environmental earth system science.)

6.4.4 By acquiring new skills for STEM PhDs for jobs

A good PhD science project gives the candidate a broad-based skills-set. The primary focus of PhD training is in designing and implementing a research project (knowing the existing literature, knowing the positions of the main players, identifying gaps in knowledge, experimental design, input of controls, replicates, statistical analysis of results), honing critical thinking skills and writing skills, whether that is communicating findings or applying for funding. Many/most PhD students also gain experience in teaching.

Twenty-first century teachers must embrace new learner-centric ways of teaching and learning. Students learn best when they are motivated to seek out new knowledge and skills. Learning is not the product of teaching; it is the product of participation and activity of the learners. The flipped classroom model is one such way of teaching which enhances student learning and achievement by reversing the traditional model of a classroom. We expect to see greater opportunities of new jobs for workers with transferable, soft skills in the wake of growing automation.

Modern skills include, besides literacy (information, media and technology) and life skills (flexibility, leadership, initiative, productivity and social), the four 'learning skills'—creativity, critical thinking, communication, and collaboration. These four learning skills need to be part of every topic in science in the same way as literacy and numeracy are. Only then can these skills be acquired by students.

We should give much greater weight to ethics when training engineers. Engineers work in a vast range of fields that pose ethical concerns. These include AI, data privacy, building construction, public health, and activity on shared environments (including indigenous communities). The decisions engineers make, if not fully thought through, can have unintended consequences—including building failures and climate change.

Machine learning, the science of getting computers to act without being explicitly programmed, has given us self-driving cars, practical speech recognition, effective web search, and a vastly improved understanding of the human genome in the past decade. Machine learning is so pervasive today that we probably use it dozens of times a day without even knowing it. Concepts of neural networks and deep learning, too, are used by modern technology to great advantage. Basics of logic are needed, similarly, to encode information in the form of logical sentences, from a computational perspective, for applications of technology of logic in mathematics, science, engineering, business, and law, etc.

Surprisingly, social studies and STEM complement each other and can work together. The lack of sidewalks, for instance, was connected to obesity, pollution, and urban planning, to name a few links. There are many more real-life problems that might be solved at the intersection of social studies and STEM [11].

6.4.5 Fusion

A non-polluting source of energy, towards which the world is moving slowly, deals with the physics of plasma confinement and stability, and is predicted to be a provider of job opportunities for STEM experts. Fusion can be tackled by an extraordinary range of innovative plasma confinement methodologies. Moreover, plasma physics and applications are so broad that the field of fusion plasmas is effectively a sub-discipline. This requires understanding of magnets in magnetic fusion energy (MFE), and lasers and optics in inertial fusion energy (IFE). Universities can join hands and push together the multidisciplinary fields required for fusion energy, while maintaining an anchor in the plasma physics required for fusion science. Universities, would thereby foster innovation, too, namely, their graduate students doing PhD need not go only for classroom instruction but can also do original research during the course of their PhD work.

6.4.6 By acquiring additional skills through attending new courses

An MBA course is in the offing in health-care innovation. It would be an eminent choice for those interested in management, innovation, and entrepreneurship in health-care. The candidates for such a course would be mostly from the health-care system, biomedicine, pharmacology, as well as those with expertize in hi-tech, research, investment, entrepreneurship, management, and public policy. Of course, an MBA degree has a strong entrepreneurial ethos for those waiting to join engaging professions.

Certificate course in data science and machine learning is another such course designed to equip professionals with existing competencies in the core focus areas, including linear algebra, statistics, gradient calculus, and programming components.

A course in mechatronics engineering would offer integration of electrical engineering, electronics, computer, and mechanical engineering. The criterion for admission to it would be a BSc degree in STEM. Mechatronics engineering would enable students to solve engineering problems between complex areas, such as using dynamic analysis of mechanical systems as a basis for the design and implementation of control systems, a systematic understanding of closed-loop control systems, and the design of control loops.

Some excellent courses are being offered free by some Ivy League colleges. STEM graduates and PhDs should keep an eye on such courses which would definitely help in getting them jobs and build their careers.

A course in computational biology, bioinformatics, and genomics can prepare students to use big-data to benefit fields ranging from ecology to medicine. Similarly, a course in computational social sciences would benefit those examining the human condition through AI during the analysis of data to understand human thoughts, beliefs, attitudes, and behaviors.

Bioethics in biomedicine is a lucrative course that needs biology and law as input subjects.

Harvard has started, in 2022, a PhD program in quantum science and engineering.

'STEAM education', with its addition of 'arts' to STEM, is a new construct. It builds upon the economic drivers which characterize STEM, with the added impetus of arts to embrace social inclusion, community participation, or sustainability agenda.

6.5 New doors/opportunities for STEM graduates, including those with disabilities

Distance education is seen as a means of developing continuing education. With digital technology, distance education has become very attractive and can acquire new markets. Open universities, which are largely public and national, play an important role by being aimed primarily at improving the skills of workers and democratizing education for the most vulnerable populations. Designed for adults already in a professional situation, an open university delivers distance education without academic admission requirements, which is precisely the definition of the word 'open' in this case. It also offers hybrid education with meeting times between students in regional centers scattered across the country.

Speech-to-text technology allows students with learning disabilities to more easily transfer their ideas onto the page. The process of vocalizing their ideas and watching their words simultaneously appear on the screen relieved much of the stress around writing. Students could watch their thoughts fill a page, proving for some that they were capable of doing so. They could then go through and revise their grammar and ideas, correcting where the technology misheard them and getting practice editing their own writing.

Engagement with both audible and visual modes of learning can also be achieved through closed captioning in class video software. Offered on both Google Meet and Zoom, closed captioning can have benefits for all students. It can make virtual classrooms that don't have sign language translators more accessible for students who are deaf and hard of hearing.

Teachers who work with students with disabilities specifically can supply their students with tools and methods of enabling accessibility technologies that they can take with them into general education classes. Students have the ability to revisit lectures, to rewind, rewatch, and take their time, as another accessibility tool. The more methods teachers offer for students to access the material and demonstrate that they've learned it, the more accessible school becomes for all students. The biggest takeaway of this online experience is just that there are things out there for free that we can use.

As a lucrative idea for skill-development amongst STEM PhD holders, researchers are exploring the microbiota as a therapeutic target for various diseases, such as metabolic disorders, cancer, and inflammatory bowel disease. A better understanding of the composition, functions, and evolutionary forces at play in people's guts and other microbiotas have potential to improve long-term health outcomes for everyone, including to tackle malnutrition or diarrhea amongst children.

Similarly, millions of workers stand to benefit from new jobs in the solar industry. Solar projects need engineers, financial experts, and skilled personnel for project development and for people with technical training for customer service.

6.6 Skills for employability in industry

While ongoing academic jobs are very difficult to obtain, PhD graduates are not well-prepared for the industry jobs outside academia. PhD graduates moving out of academia need to re-train themselves because there is a sharp contrast between university and non-university occupations in terms of workplace cultures and employer expectations. For example, industry employers want skills needed for work rather than qualifications or publications.

For doctoral students to train suitably, career consultations from both universities' career centers and industry experts should be offered early in PhD programs to help students make informed decisions about future options. Universities should strengthen their partnerships with industry to facilitate work experience. Those seeking academic jobs also need to be provided with meaningful opportunities to work alongside academic staff in both teaching activities and research projects.

Two-thirds of PhD students in STEM fields (science, technology, engineering, and maths) would be working in industry. The banking, civil engineering, mining, energy, and medical/pharmaceutical sectors are the top employers of PhDs. Out of over 20 000 PhDs recipients in the US per year, the industry currently absorbs about half of them for jobs, and this number keeps increasing every year.

Universities need to ensure that during doctoral programs students are guided for employment possibilities inside and outside academia. This includes opportunities to build transferable skills such as teamwork, communication, analytical skills, and leadership. This also means teaching students how to write and speak for different audiences beyond academia, including policymakers and the public [12, 13].

Due to the rapidly changing nature of work, employment in jobs requiring STEM skills is growing faster than in other jobs.

Sectors including health-care and social assistance, education and training, construction, and customer service, would all require strong numeracy and computational thinking skills, including problem solving.

The importance of human skills in automated industries cannot be under-stated in interpersonal and creative roles, where uniquely human skills like creativity, customer service, care for others, and collaboration are required regularly.

A large number of employers, and this number is increasing, consider employability skills to be at least as important as technical skills. Those looking for jobs need to emphasize their employability skills, rather than just the technical skills. Communication, reliability, teamwork, patience, resilience, and initiative are required for all jobs, and this will continue to be the case in the future.

Broadly speaking, some of the skills needed are as follows:

1. Cognitive flexibility: the ability to adapt to the changing world and information around you.
2. Lifelong learning capacity in traditional and digital literacy.
3. Basic literacy, numeracy, and media literacy (including the use of technology).
4. Creativity and imagination: the human traits that separate us from machines and bring a human perspective to our work.
5. Computational thinking: problem solving processes.
6. Ethical and sustainable practice: a commitment to do no harm to humans or to the planet.
7. Indigenous perspectives and cultural competence: promoting reconciliation and working successfully and respectfully across cultures and customs.
8. Well-being: taking care of our minds, and bodies.

A further asset of a STEM PhD can be the computational skills, which are the ability to understand a complex problem, develop possible solutions, and then present these solutions in a way a computer, human, or both, can understand. These skills are what primary math should aim toward, emphasizing interdisciplinary connections across key learning areas.

Strong basic numeracy skills build a foundation for a lifetime. These skills are taught across the disciplines, including science, geography, visual arts, health and physical education, languages, history, and design.

There are major challenges of training and research on interdisciplinary approaches which are still too little explored although they seem urgent.

Vocational education and training (VET) is where you learn skills for employment. People who have a VET qualification and work in the agricultural, forestry and fishing, or mining industries have similar, if not higher, weekly earnings compared to those who have a university degree.

6.7 Apprenticeships and training in industry, for facilitating employment

Paid internships, a vital recruiting channel in technology, are scarce, and STEM PhD students should keep an eye on them.

A robust program of internships/apprenticeship is run by Pratt & Whitney at its large 1.2 million-square-foot factory in North Berwick, Maine (USA), where it makes engine parts for passenger airplanes and the F35 joint strike fighter. They need 150 new entry-level workers at the site. These new hires spend six weeks in initial training and as much as another six months learning on the job to run machines the company says cost as much as $1.5 million each, which mill and grind parts to specifications within the thousandths of an inch. The company also runs a 3-year apprenticeship program for prospective managers in collaboration with nearby York County Community College through which participants earn associate degrees and get 8000 h of training on the job. Pratt & Whitney has similar partnerships with community colleges near its facilities in Florida, Connecticut and Georgia [14].

References

[1] Smye S W and Frangi A F 2021 Interdisciplinary research: shaping the healthcare of the future *Future Healthc. J.* **8** e218–23

[2] Casassus B 2023 France's research minister has a plan to shake up science *Nature* **March 2023** https://doi.org/10.1038/d41586-023-00628-7

[3] FUN 2023 In 2024, France Université Numérique is launching a new offer aimed at working people https://fun-mooc.fr/fr/actualites/en-2024-france-universite-numerique-lance-une-nouvelle-offre-des/

[4] Conroy G 2023 Can India's new billion-dollar funding agency boost research? *Nature* **619** 681–2

[5] Golden Gate University 2024 Doctorate in business administration in emerging technologies, with specialization in generative AI https://upgrad.com/dba-emerging-technologies-specialization-in-gen-ai-ggu/

[6] Melissa E 2024 Germany, once a powerhouse, is at an economic 'standstill' *New York Times* https://nytimes.com/2024/01/18/world/europe/german-economy-standstill.html

[7] Foster P, Gross A and Borrett A 2023 The looming financial crisis at UK universities *Financial Times* https://ft.com/content/0aca64a4-5ddc-43f8-9bba-fc5d5aa9311d

[8] Leszczynska A 2023 The IAEA's commitment to transforming lives through inclusive education *IAEA* https://iaea.org/newscenter/news/the-iaeas-commitment-to-transforming-lives-through-inclusive-education

[9] Maliniak D 2023 How do you stay on top of technology *Microwaves & RF* https://mwrf.com/resources/industry-insights/article/21279217/microwaves-rf-how-do-you-stay-on-top-of-technology

[10] Stanford Woods Institute for the Environment 2021 Inspiring collaboration: grants empower experts to tackle environmental challenges https://woods.stanford.edu/news/inspiring-collaboration-grants-empower-experts-tackle-environmental-challenges

[11] Fultz A 2021 Combining social studies and STEM in a project-based learning unit *Edutopia* https://edutopia.org/article/combining-social-studies-and-stem-project-based-learning-unit

[12] Victoria University 2023 Australia has way more PhD graduates than academic jobs. Here's how to rethink doctoral degrees https://vu.edu.au/about-vu/news-events/news/australia-has-way-more-phd-graduates-than-academic-jobs-heres-how-to-rethink-doctoral-degrees

[13] Clancy M *et al* 2023 To speed scientific progress, understand how science policy works *Nature* **620** 724–6

[14] Marcus J 2021 A surprise for America's many career switchers: they need to go back to school *Hechinger Report* https://hechingerreport.org/a-surprise-for-americas-many-career-switchers-they-need-to-go-back-to-school/

Employability for PhD Students in STEM

Jatinder Vir Yakhmi

Chapter 7

Overcoming disruptions against career growth

We shall discuss different situations which disrupt a career, especially the career of a PhD holder in STEM. Among the examples discussed are: how the pandemic, a sudden disruption, taught us to adopt new learning processes and new professions; how does one handle sudden loss of a job; or the adverse influence of a personal sickness, or loss of a family member, in mid-career; and, as a professional researcher how to stay clear of plagiarism, which can adversely affect your career.

7.1 Employment and career options for STEM PhDs

Doctorate programs provide an opportunity to young minds to tackle intellectual problems and explore new areas of knowledge. Getting a doctorate is intellectually rewarding, but it is not financially rewarding, at least not in the short term.

A majority of PhD students start out with aspirations to become professor, but only a tiny fraction of PhD graduates actually succeeds in getting a tenure job in academe. During the recent past, the number of young aspirants finishing PhD has gone up substantially, but the number of faculty positions has stagnated.

Post-PhD, many researchers leave their academic research groups and don't pursue a postdoc there, gotten tired of the publish or perish grind, the short-term contracts, the toxic egos, the countless additional responsibilities, the strain on the family, the administrative burden, and the mediocre salaries during PhD work, or later as a postdoc.

Then, there is the divide between basic and applied research. Basic research results in general knowledge and an understanding of nature and its laws. This general knowledge provides the means of answering a large number of important practical problems, though it may not give a complete specific answer to any one of them. It is left to the applied research to provide such complete answers. Yet the progress of industrial development would eventually stagnate if basic scientific research was neglected. New products and new processes are founded on new principles and new concepts, which in turn are painstakingly developed by research in the purest realms of science.

It is a myth that the skills of a STEM PhD are so niche that they would struggle in other professions. On the contrary, during their period of doctoral work a STEM candidate trains themselves as a mentor, a people manager, a project planner and manager, a teacher, a multi-tasker extraordinaire, a writer, an illustrator, a public speaker, a budgets manager and writer of business plans, and finally an academic researcher, an author with exquisite attention to detail and the ability to distill complex ideas into simple messages. The discovery of new therapeutic agents and methods, for instance, usually result from basic studies in medicine. Of course, for the fruits of this research in medicine to reach medical practitioners requires teamwork involving the medical schools, the science departments of universities, government, and the pharmaceutical industry.

7.1.1 Jobs in industry

Only about 20% of postdocs will land in tenured or long-term academic positions. Some other posts can be considered successful outcomes too, including those at minority-serving institutions, liberal-arts colleges, community colleges, non-profit research institutions, policy think-tanks, biotechnology and pharmaceutical companies, and other types of industry.

It is a struggle for postdoc academics to secure a job in 'industry', and the regular job-hunting leads to insecurity and desperation, considering that they have spent the best years of their life gathering credentials and prestigious academic achievements.

Statistically speaking, 10 years after receiving their PhD, about 85% of graduates are directly engaged in academic or industrial research, usually after a period as a postdoc; 10% work in non-research activities related to science; and 5% opt for careers unrelated to science.

Many aspiring female engineers in Japan choose non-STEM jobs due to the social stigma that women in the STEM field are too busy at work to attend to families, so have a hard time finding husbands. In the IT field alone, Japan is looking at a huge shortfall of workers by 2030, largely due to a severe under-representation of women. That results in a decline in innovation, productivity and competitiveness for the world's third-largest economy. To close the gap, the Japanese government is planning over 100 STEM workshops and events mainly targeting female students. Companies like Mitsubishi Heavy Industries and Toyota are offering scholarships to female STEM students to attract talent.

STEM PhDs do look for positions in industry, which means full-time jobs in companies, non-profit organizations and government agencies.

We should also take note that subjects like design and technology, music, art, and drama are important for children to develop imagination and resourcefulness, resilience, problem solving, team-working, and technical skills. These are the skills which will enable young people to navigate the changing workplace of the future and stay ahead of the robots, not exam grades.

7.1.2 Transferable skills of value in business companies

The occupational landscape of STEM fields is complex and continually evolving. Advances in scientific research and subsequent applications to business and industry will result in new occupations and new demands for specific STEM related skill sets. Candidates pursuing STEM fields must be equipped with not just STEM-specific skills and knowledge, but also career adaptability, to be able to navigate any change with self-knowledge, confidence, and planning.

STEM is more than a simple sum of four disciplines. Within the traditional STEM domains, there is a great variety of subdisciplines that warrant separate consideration. Also, the boundary of defined STEM fields has been expanding to include non-traditional areas, such as arts, medicine, and even managerial or sales positions.

As a research scientist, the irony is that a STEM PhD is highly over-credentialed, as well as underqualified for most jobs in industry. In some industry jobs like project management, you might get away with selling your transferable skills.

STEM PhD students do not routinely receive such tailored training in the practical professional skills that are highly valued in the business sector, such as negotiation, communication, business strategy, basic economics, and marketing, as abundantly represented in the curricula of leading MBA programs.

In the meantime, this training gap can easily be filled by taking short courses or programs in business skills. Many are available online as massive open online courses, and some offer certifications.

It is also useful to take courses in data science, machine learning, and natural-language processing.

7.1.2.1 STEM PhDs do have transferable skills useful to the business world
Career coaches for industry jobs ask you to highlight all the useful 'transferable skills' you acquired as an academic, such as data analysis, critical thinking, project management, leadership, effective verbal and written communication, and so on.

If you have earned a science PhD, you were probably told by mentors, advisers, and career-development specialists that you will need to develop a lot of new skills to succeed in any sector outside academia. But your PhD program has already conferred many skills that are important, even crucial, in the business world, and that are comparable to—and in some cases superior to—the talents acquired in a graduate-level business program. Here are some examples:

Data analysis: During your STEM PhD program, you were trained to gather, evaluate, synthesize, and present data, and to uncover relationships, correlations, and trends. The business world increasingly relies on the same methodologies to develop strategies and identify opportunities.

Resourcefulness: You probably had to create experiments, methodologies, and analyses with limited resources and under tight time constraints. Successful business people are often challenged to develop a product or service while facing the same difficulties.

Technological awareness: You were trained to understand the fundamentals of a range of technologies. Many of these technologies are at the heart of products and services in the private sector.

Resilience: You may have encountered unexpected setbacks in your research or studies, yet powered through to reach your goals. This resilience in the face of challenge often separates the most successful entrepreneurs from the rest.

Project management: Completing a PhD typically requires the coordination and scheduling of disparate resources and individuals—as well as thinking through all aspects of a complex project or activity. The same course of action is a core component of the business world.

Problem solving: You had to use novel thinking and innovative frames of reference to identify and solve technical problems. The ability to reframe problems to identify novel solutions is a key skill in business.

English proficiency: You are well-skilled in English, the language of international business.

Written communication: PhD holders have good experience in writing and describing complex ideas and methodologies. Effective written communication is crucial to business success.

It is easy to gauge that a STEM PhD does possess most of above-mentioned transferable skills.

7.2 Disruption due to the Covid-19 pandemic, or other disasters: disaster-proofing

The Covid-19 crisis taught us to adapt. People, after the coronavirus epidemic, have realized that your time is limited and you could lose anybody at any time. The Covid-19 crisis had much of the population hunkering down at home, leaving many companies and non-profits figuring out how to adjust to remote work for an uncertain period.

The pandemic, by pushing work, education, and many other aspects of daily life onto online platforms, has magnified the technological shortcomings of even countries advanced in high tech.

Many established research scholars were thrown off-track vis-à-vis where they had hoped to be before Covid-19 set in. Many new scholars lost the benefit of substantial in-person, live mentoring.

During the three years of Covid-19, millions of students left college/university. Added to the prevailing uncertainty, this exodus was also caused by an already existing undercurrent of skepticism about the worth of a college degree versus the high costs to get it. Even when financial aid is counted, the inflation-adjusted average cost of a 4-year college education has more than doubled since the 1970s.

Facing a dizzying array of pandemic tragedies, social movements, and inequities, college students created new norms, such as low in-person attendance and disengagement, and all that continues still, even if not to that large an extent.

The pandemic has taught students that they can get most of the course content by reading the textbook or watching a recorded lecture. So, what is the value of coming to class? We need to rediscover the unique advantages of learning together in a shared space. It has always been human connection—the give and take of discussion, the knowledge gained through solving a problem together, the fun of exploring an outlandish counterfactual, or the sincere inquiry of a spur-of-the-moment question. In essence, connecting with students on a human level [1].

7.2.1 Covid-19 and workplace changes

The pandemic blurred the lines between spaces—one's home, workplace, and the childcare centre, all converged into a single, tight space. One impact of this was a lack of energy, motivation, and time to write research proposals to secure new funding. This has brought focus on the need to implement family-friendly policies, such as allowing mothers to bring their kids to work when childcare is unavailable.

The Covid-19 pandemic made the workplace change beyond recognition. Offices gave way to homes; conference rooms were abandoned and Zoom took over. As 2022 began, organizations began talking about hybrid models of work, where employees were asked to come into offices for a few days every week. As 2022 ended, the hybrid model began giving way to a 'back to the office' model, returning to the old, pre-pandemic ways of working. But many employees who have experienced the benefits of flexible working demand flexibility about working from anywhere, claiming that flexible working could lead to increased worker happiness and even a higher productivity.

7.2.2 Mothers affected most among staff during Covid-19

The Covid-19 pandemic has had a disproportionate impact on mothers in academic science. Universities throughout the world failed to acknowledge that they perform a 24-h job to take care of kids at home. Consequently, during Covid-19 times, scientist mothers were significantly less likely to submit manuscripts and to meet deadlines than other researchers, leading to fewer professional opportunities. Some mothers had to even leave the workforce altogether.

Many scientist mums still struggle to find the support needed to progress their careers. Moms have to leave to pick up children from childcare, are not able to travel to conferences and have less time for their work. In science, time to concentrate is crucial, because you can't write good papers, do data analysis, or write proposals without it. During the Covid-19 lockdowns, that time just vanished. Lockdown provided an opportunity for working mothers to bond with their children and get more involved in their education, sometimes even sharing a laptop, when children had to join a Zoom class for school, even if that meant catching up with own work at night.

Academic institutions, publishers, scientific societies, and funding agencies ought to address inequities faced by scientist mothers and provide better support through policy changes and practical mechanisms, during and after a pandemic-like disaster, in particular.

Universities should run mum-friendly facilities that give them a place to bond with their children and thereby work effectively.

7.2.3 Economic anxiety in the wake of Covid-19, and non-lab positions

Apart from affecting the health of people, the pandemic also brought economic anxiety. A huge number of workers lost their jobs, and even lost their rented accommodations. Feeding their families became a big challenge. The Covid-19 pandemic opened the eyes of STEM workers to look for positions which are not lab-based because it gave them flexibility to work anywhere. In general, PhDs and less experienced postdocs entering industry often start out as member of a team focused on a particular research area. They can go on to become senior scientists, and, typically after 3–5 years, principal scientists. This is roughly equivalent to the position of a group-leader in academia. Those with more postdoctoral and work experience before joining industry, can sometimes start at the principal scientist/investigator level, with managerial responsibilities.

7.2.4 Wars

Science has an ingrained international character, because researchers recognize the importance of maintaining the free flow of knowledge even during conflicts like a war. But wars tend to change those priorities of extending funds for supporting science. Wars in Ukraine and Israel are already affecting research, and could change it for years or maybe decades in the case of some affected labs.

7.2.5 Disaster-proofing

To mitigate risk to personnel and to precious lab equipment, one needs to remain in readiness and prepared for sudden natural disasters (earthquake, hurricane, tsunami, cyclone, thunderstorm). To retain one's lab equipment (and thereby the job), it might be essential for a STEM graduate to switch off equipment not in use, and ensure that any key equipment required to be kept ON is kept at table tops, to avoid being ruined by floodwaters. Most emergencies in research labs arise from water leaks and electrical power interruptions. Switch gears controlling the generators are often located in the basement, which flood heavily, knocking out power. When freezers thaw, research animals die and precious research material becomes waste.

7.3 Lay-offs, quiet hiring

7.3.1 Lay-offs

2022 was the worst year that the tech industry had experienced in recent times. Apple, Amazon, Alphabet, Microsoft, Meta, and other high tech companies lost a combined $3.9 trillion in market value. Several companies announced austerity measures in 2023 to cut costs, including plans to lay off thousands of workers. LinkedIn, owned by Microsoft, said it will lay off hundreds of employees worldwide, including teams dedicated to engineering and marketing in China, because of slump in demand.

According to Layoffs.fyi, the tech sector lost around 150 000 jobs in 2022 (including at Meta, Twitter, and Uber), and suffered more than 50 000 cuts in 2023.

Google laid off thousands of employees in January 2023. Google is not alone. Dozens of companies employing large number of employees have laid off employees en masse, to save money when the business is low. Restructuring in many existing industries is leading to lay-offs in thousands while a future in which new projects could be driven largely by automation and robots could put paid to the aspirations of millions of young men and women readying to join the workforce every year.

The shame of having lost one's job can be devastating. It can lead to anxiety, sadness, anger, and frustration. Being laid off should not be a judgment on who you are. You are not your job or your earning potential. Who knows, a lay-off might turn out to be a shift to something better.

After cutting costs, many once-promising tech companies have been facing closure [2].

A shrinking big tech job market is a new reality that is setting in for recent graduates who spent years honing themselves for careers at the largest tech companies.

2023 has been the most difficult year for start-ups in at least a decade. Approximately 3200 private venture-backed US companies have gone out of business. Venture capital firms are deciding which young companies are worth saving and urging others to shut down or sell, urging their founders to walk away from doomed companies.

Private 'unicorn' companies worth $1 billion or more had seen better days. In 2023, several 'unicorn' companies were cutting costs, laying off staff, and otherwise adapting to the difficult market.

7.3.1.1 Advice to individual job seekers in the face of lay-offs
If you accept a new job, do plan to start that job before resigning your old one with vacation time at your old job. Make sure that new job still exists and it is what you were told about, keeping two weeks with you before resigning from the previous job. Resignations are usually followed by an immediate walk-out and loss of credentials, so any loss caused by a quick resignation at the old firm should not make you totally jobless.

Job cuts can have an immediate effect on wellness. Being laid off increases the risk of a host of health conditions. Therefore, the best time to look for a new job is when you've already got one. You won't feel desperate, and you can take time and make the best decision.

Businesses in several industries which are in serious financial troubles are even rescinding job offers they made just a few months ago. Lay-offs and downsizings are an unfortunate reality of corporate life. But rescinding job offers can be harsh, especially for college graduates who may have gotten a job offer early in the recruiting season, and hence missed out on further interviewing, or for people who made significant life changes like selling a house or relocating on the basis of an accepted job offer. Therefore, the lesson to be learnt by job-aspirants, including STEM PhDs is never to leave a job until you have another lined up, and that does not imply just an offer, which is only an assurance, and not a secure job.

Never stop looking for a job if you are unemployed. Continue honing your skills, learning new ones or working and expanding your network. Insufficient technical ability is the biggest reason for a no-hire. When a job offer *wasn't* extended, by far the main reason was lack of technical ability. You always should have a Plan B for good and bad times.

Neither self-blame nor blame others for your lay-off, it won't be helpful in the long run. Instead, think how you can move on after what has happened. Stay positive during low-times, and have faith in friends and mentors who truly want to help you. All you have to do is ask.

Despite signs of economic recovery, including lower inflation rates and sustained low unemployment, 2024 has already seen a surge in lay-offs, surpassing 10 000 in the tech industry alone. The technology industry continues to experience lay-offs in 2024. Over 34 000 employees already lost their jobs across major companies like Google, Amazon, and more than 100 other tech firms in the first few weeks of 2024.

Big tech companies are spending billions of dollars on the expensive chips and supercomputers necessary to train and build AI systems, and AI technology that they believe could one day be worth trillions. By the end of 2024, Meta is likely to have purchased a huge number of specialized chips, each costing an estimated $30 000 from the chip maker Nvidia. High tech companies are trying to shed expenses by laying off some employees to pay for its growing investment in AI. Google laid off hundreds of workers to lower expenses as they invest heavily in AI. Meta, Amazon, Microsoft, Google, and Apple have cut about 112 000 jobs from their respective peaks in 2021 and 2022.

How should one avoid a lay-off? Technologies of generative AI and large language models (LLMs) such as ChatGPT are predicted to replace human workers at a large scale, first among them are those who don't have AI skills. Employees must advance their hard skills, such as becoming technically proficient in AI applications or new programming languages. Those not doing so may be targeted for lay-offs.

To safeguard against a potential lay-off, it's critical to stay informed about market trends and assess whether your chosen career is prone to offshoring or automation. Managers who lack hands-on involvement and not directly contributing to task and project execution might be considered expendable during cost-cutting measures. Hence, do not be a quiet invisible, lest it may bring your downfall and lay-off.

7.3.2 Quiet hiring

Quiet hiring is being resorted to by organizations to facilitate acquiring of new skills and capabilities by their existing employees, without adding new full-time employees. Under quiet hiring, they are bringing a focus on internal talent mobility to ensure employees address the priorities that matter most without changes in headcount, upskill existing employees, and leverage alumni networks and gig workers to bring in talent as and when needed.

7.4 Disabled

7.4.1 Hurdles galore for the disabled

It is heartening to note that early-career scientists who face physical challenges such as blindness, deafness, or paralysis are getting employed, even though only in small numbers, with varied and rewarding careers. They work in academic, government, and industrial research as teachers, consultants, etc.

Finding work is not easy for disabled researchers who have trouble seeing or hearing or who have limited mobility. They need creative workarounds in the lab and field. For success in getting employment, they need grit and ingenuity. They may have to design their own equipment. Besides, they need a crew of friends, peers, and mentors who can provide support. And the work they seek must capitalize on their strengths.

Researchers who have a degree but also a disability and want to work in a scientific field must first ensure that physical adjustments are made to labs and other workplaces to facilitate access. These can include redesigning lab sinks to accommodate a wheelchair, or checking that doorways to halls with lifts or ramps don't lock automatically and block exits. Visually challenged candidates require posting of emergency instructions in Braille.

Many labs remain difficult or impossible to navigate for scientists who have mobility issues: aisles are too narrow, tables are too tall, and eyewash stations are tucked into inconvenient corners. Physical barriers are not the only obstacles, bias can also be an issue for those using a wheelchair after sustaining a spinal injury. Disability will make it hard for them to find a job. Even the most basic research activities—accessing the literature, submitting papers, attending conferences, and reviewing manuscripts—require time-intensive, personalized workarounds, not easy for blind researchers, in particular.

There are practical hurdles to using assistive tools. Even when Braille versions of textbooks are available, teachers may not know about them or be able to order them in time for a class. Figures, tables, and graphs typically aren't translated into Braille, so a student with a visual impairment often needs to collaborate with a sighted colleague to interpret visual data.

7.4.2 Innovative new solutions for disabled

Science provided solutions to such calamities as famine and plague, transforming them from incomprehensible and uncontrollable forces of nature into manageable challenges. Merit based science is *truly* fair and inclusive. It provides a ladder of opportunity and a fair chance of success for those possessing the necessary skills or talents. Some scientists with disabilities have reframed their impairment as a positive attribute: they say that coping with the challenges of everyday life has helped them to develop unusual skills and expertise. Navigating the town can train their brain to make mental maps, on the spot. Similar thinking can help them in organic chemistry, or other experimental labs.

Scientists who have disabilities use a growing array of specialized equipment to enable them to carry out research. But many of them find ingenious ways to make things work. Scientists with disabilities are installing light switches at an accessible height, as well as which are labeled for people who rely on Braille. Lab benches need to be height-adjustable [3].

Sonification provides a solution for scientists with visual limitations to 'see' data, for instance in astronomy. By translating numerical values into sounds with certain parameters—for example, a star's brightness might be encoded as pitch—researchers can home in on important changes. Researchers in genomics and geology are also exploring sonification, although protocols that would allow scientists to compare data in auditory formats are still under development.

Produced from plastic thin enough for light to shine through, lithophanes can encode multiple forms of chemical and life-science data, examples of which are a scanning electron micrograph of a butterfly wing, the bands of an electrophoresis gel, or the ultraviolet spectrum of a protein.

The need for accessibility by the disabled researchers in science is growing, and new tools are on the way. A data visualization tool, called the multimodal access and interactive data representation system (MAIDR), encodes data as both sounds—termed sonification—and Braille, providing tactile analysis with the help of a refreshable Braille display. They can hear the trend of the data in sound, and feel the pattern!

As discussed earlier in chapter 4, if not addressed promptly, the flaws in the working culture of AI can perpetuate biases that affect the resulting technologies. For instance, using AI to help diagnose diseases can transform people's lives, but it may not help remote areas like Africa, if data from Africa are not collected to understand African health care and related social-support systems, sicknesses, and the environment people live in.

AI conferences ought to be inclusive and accessible to disabled people, to let them attend. Many disabled researchers have concerns about the barriers they face in AI. There is a need for teaching them machine learning, data science, and AI.

7.5 Debt ceiling

Top colleges in the US charge upwards of $70k per year for college. Most students in USA have no financial resources for college studies and they take loans, tens of thousands of dollars each, to be paid back later, in installments. Most undergrad students are 17 or 18 when they sign onto such a large amount of debt. A debt burden on teenagers is a moral hazard. Under a forgiveness plan, the Biden administration has canceled $136b in debt from over 3.7 million students in USA, whose family incomes were less than $75k per year.

Some people argue, however, that it is not fair to provide debt relief to some individuals while others have already fulfilled their obligations. They suggest not to write off this student debt, but make the payback manageable and fix the system going forward. Alternative schemes, such as improving access to affordable education or addressing the rising costs of higher education, should be more effective in addressing the underlying issues.

7.6 Bias towards gender, black, Asian, candidates coming under adult education

7.6.1 Gender bias

7.6.1.1 General

Science, technology, engineering, and mathematics (STEM) skills and the industries they power are in increasing demand and are opening new areas of economic activity. In the next decade there will be even more demand for engineering, more investment into scientific research and rapid development of technologies such as AI, quantum technology, robotics, automation, precision manufacturing, and advanced materials. These fields will also drive the growth and emergence of other high-value, high employment industries such as fintech, medtech, agritech, cybersecurity, clean energy, renewables, and other green technologies. To meet this economic opportunity, the world must adapt and carry out full attraction, retention, and advancement of people in STEM careers.

For all demographic groups of STEM workers in USA, those with a bachelor's degree or higher have higher median earnings than those without college degrees. Median wage and salary earnings are higher for those working in STEM than in non-STEM occupations, regardless of sex, race, ethnicity, or disability status. Additionally, within the STEM workforce, higher education, like a PhD, translates into higher pay.

STEM industries are widely acknowledged as growth sectors. But women hold a very low percentage of leadership positions across the globe, viz. fewer than 10% of leadership positions in STEM related industries in Australia. This needs to change.

7.6.1.1.1 Women are overlooked in STEM

Women in STEM are overlooked for their technical expertise, and ignored and interrupted when they offer that expertise. A 2018 study by the Pew Research Centre noted that while 74% of women in computer jobs in the United States reported experiencing workplace discrimination, 62% of African–American employees also reported racially motivated discrimination. Women of color in tech find themselves doubly affected.

Although open recruitment with no gender bias is expected to be implemented for jobs in STEM subjects, the proportion of female applicants is abysmally low, due to systemic bias in recruitment practices—including a perceived bias against female candidates in open recruitment. The under-representation of women in STEM has been the subject of widespread debate. Let future applicants have an assurance that the universities would henceforth support gender equity, diversity, and inclusion.

Due to cultural norms of masculinity STEM fields can be hostile to women. STEM environments are often hostile to LGBTQ+ people (lesbian, gay, bisexual, transgender, and other sexual and gender minorities), leading to under-representation of these communities.

7.6.1.1.2 Women with PhD in STEM

It is the problems with workplace culture that makes women leave STEM jobs. More women with tenure-track and tenured academic posts leave than their male counterparts. The retention gap between men and women begins to increase about 15 years after academics finish their PhDs. At that point, many of the faculty members would be expected to have received tenure. Women feel pushed out [4].

Quite often, families steer women away from STEM careers based on the notion that women in the STEM field are too busy at work to juggle families, so have a hard time finding husbands.

Japan is looking at a shortfall of 790 000 IT workers, by 2030, largely due to a severe under-representation of women [5]. An 'unconscious bias' deters girls from pursuing STEM studies. If we don't have a gender balance, then technology is going to suffer. Japan has only 16% of female university students majoring in engineering, manufacturing, and construction, and that too, with just one female scientist for every seven. That is despite Japanese girls scoring second-highest in the world in maths and third in science!

The percentage of female college students in USA majoring in engineering has grown markedly from 1970 to 2020, but of that fewer than 20% engineering majors and with bachelor's degrees are women. The same is true for the technology sector, where three-quarters of all computer science graduates are male.

The number of women in management and leadership roles thins at every stage of the career pipeline. In Australia, women make up just 28% of management positions in STEM, and only account for 8% of CEOs and heads of businesses.

Encouraging signs are that in 2020, US women were awarded approximately 17 000 doctoral degrees in S&E fields, which showed an increase by 18% over the decade 2011–2020. Women earned 53% of doctorate level agricultural and biological sciences degrees. In physical and earth sciences, women earned 34% of doctoral degrees in 2020. Physics remains a key field where women are significantly underrepresented among degree recipients.

7.6.1.1.3 Bias against middle-aged women

Male STEM workers typically get higher remuneration than female STEM workers regardless of whether they have an advanced degree. Women made up 51% of the total labor force in USA with at least a bachelor's degree in 2021.

Evaluations of male professors remain generally consistent over time, while women experience a quick decline from their initial peak in their 30s and hit rock bottom around age 47. If one looks at both gender and age together, then one can see better how the perceptions of men and women differ. Feeling less capable is a struggle every woman who has sat through a meeting with overconfident men already knows. As women move into positions of evaluating others, they should speak up about double standards and be agents of change away from these bias-laden practices.

Meritocracy should be the only benchmark for progressing up the career ladder, with no gender discrimination.

7.6.1.1.4 Patents by women

Patents provide a 20-year monopoly over a new invention and are a well-known measure of the output from STEM-based industries.

Getting a patent can be important for career progression and for securing investment capital. The number of patent applications from female inventors, has grown over the past 20 years. What has been less clear is whether these applications convert to granted patents.

Among over 30 technical fields, counted as categories recognized for patents in science, a majority of female inventors are clustered in just a few of them, i.e. in the life sciences, chemistry, biotechnology, and pharmaceuticals and medical technology.

7.6.1.1.5 How to rectify the bias against gender

Women in STEM often experience hostile work environments, including bullying and harassment. STEM organizations need more clear pathways for women in untapped talent pools to return to or move into STEM leadership roles. STEM organizations must develop mechanisms to re-skill, re-integrate, and retain women over the long-term. They should offer training, re-skilling, and return-to-work programs for their existing talent—whether with or without technical work experience.

A barrier to gender equality in STEM is partly rooted in the expectation that women are the primary caregivers and a perception that caregiving will detract from a woman's ability if she is given greater leadership responsibility.

7.6.1.2 Mothers in science

A coalition of organizations around the globe that collectively represent millions of women in science, technology, engineering, mathematics, and medicine (STEMM), led by Mothers in Science (MIS)—an international non-profit organization based in France—released a report recently that outlines policies for funders that, it says, will banish the long-standing discrimination against scientist mums and overall gender bias in the scientific enterprise.

In 2021, MIS held a conference to bring together groups studying gender discrimination in STEMM and to share the preliminary results of their global survey, which reached roughly 9000 researcher-respondents in 128 countries, including parents and those without children. The report highlights six focus areas. These include: (i) the need for financial support to ensure research continuity; (ii) flexibility for parents and caregivers, including remote working options; (iii) systems for tracking diversity and inclusion; (iv) system for flagging suspected discrimination; (v) a simplification of the application and evaluation process for grants and fellowships; and (vi) addressing the disproportionate impact that the SARS-CoV-2 pandemic has had on female scientists.

If we want progress and excellence in science, then we should all be focussing on inclusivity, which includes women and mothers. There's no way around it.

7.6.1.3 Women in STEM in Australia

Of the STEM qualified population in Australia, women comprised only 17% in 2016 [6]. For traditionally female dominated fields such as biology—women comprised 56% of postdoctoral biology academics in 2016, but only 18% of professors. These issues cut across much of the STEM scenario. In engineering, women represented only 12.4% of the workforce in 2016, with men more likely to be employed at higher levels of responsibility and women at less senior levels. Lack of support networks, including mentors, career sponsors, and professional groups, contributes to women feeling out of place in STEM fields.

7.6.1.4 STEM recharge project: UK women can return to STEM careers
Under a 'STEM Recharge Project', around 75 000 scientists in the UK, mostly women, are currently looking to return to the science and engineering sectors after taking career breaks due to caring responsibilities. The UK government has awarded £150 000 to support a pilot scheme that will help parents and carers return to the science sector after career breaks. The STEM Recharge Project will offer free career coaching, mentoring, and sector-specific upskilling to 100 returners who have taken career breaks of at least one year. The STEM Recharge Project will be delivered in the midlands and north of England and will provide support for people returning to the workforce [7].

7.6.2 Bias towards black and Asian workers

In USA, most professors in science are white and male [8]. Progression of doctorate candidates up the seniority ladder is based on being vouched for by other academics, as well as access to information on grants and how to win them. Senior academics usually see more potential in white people. The lack of diversity feeds inequality, because those having authority hardly ever empathize with problems that black and minority ethnic researchers face. Funding discrepancies are a major concern. Researchers of marginalized ethnicities rarely win grants they apply for, when compared with their white peers.

Innovation and disruptive technologies are salient components of industrialized economies. Robotics, 3-D printing, precision machining, data analytics, bioinformatics, digital imaging, design and animation all feature prominently in this. Global companies usually employ workers with productive capacities that can be used in manufacturing, science, and technology-intensive sectors, as well as information technologies. Unfortunately, not many young candidates from African countries, e.g. Nigeria, possess these frontier skills. Without addressing the problems of skills mismatch and the lack of digital skills, young Nigerians (or other Africans) will continue to miss out on opportunities of jobs in large global corporations.

Quite similar to the bias against black communities, there also exists a bias towards Asian non-whites.

7.6.3 Adult education

Adult education is a gift. It changes lives. But opportunities for adult education have been shrinking, across the world. Not everyone grows up in an environment full of options. Something has to change, and rapidly, or else uneducated adults will never have opportunities.

7.7 Quiet quitting, working from home, job-switching, great resignation

7.7.1 Quiet quitting

It's great to have a job you're passionate about. But some workers are walking away from dream roles, in search of stability and security. Individuals have been pushed so hard for so long, that apathy sets in, motivations wane and people are exhausted. No more bringing work home and perpetuating the imbalance between work and home life. You're still performing your duties, but you're no longer subscribing to the hustle-culture mentality that work has to be your life. 'Quiet quitters' are people who have grown disillusioned with their workplaces and given up putting in additional effort.

If you're unhappy at work, but leaving your job isn't an option, then you may want to try '*quiet quitting*'. This trend of simply doing the bare minimum expected at work has resonated with young people. Tired, overworked, burnt-out employees are refusing the working conditions that are unsuitable. Quiet quitting is the term used when workers only do the job that they're being paid to do, without taking on any extra duties, or participating in extracurriculars at work.

Quiet quitting refers to opting out of tasks beyond one's assigned duties, implying no more staying late, showing up early, or attending non-mandatory meetings, without subjecting their teams to an unsustainable 'hustle' culture. It's just doing your job and acting your wage to meet your financial needs and have time to drive your kids to and from school, and to take a day off, when needed.

Disappointed young people are increasingly seeking flexibility and purpose in their work, and balance and satisfaction in their lives. Many young professionals are now rejecting the live-to-work lifestyle, by continuing to work at minimal capacity, but not allowing work to control them. Initial disregard for financial security is something among some workers who are looking for a more fulfilling career.

Since Covid-19, the young generation does not wish to continue working 24/7, which meant being always available. They are still performing their duties, but are no longer subscribing to the hustle mentality, realizing that work is not their life, or their worth as a person. They have realized that other than burnout and quiet quitting, there is also a third alternative—being engaged in their work without burning out and sacrificing their health and happiness.

'Quiet quitting' has struck a nerve. It means more time for friends, family, and personal pursuits, not to mention a side hustle. But this workplace trend has drawbacks. Despite what the name suggests, quiet quitting doesn't mean turning in a resignation letter. Instead, it's a stealth retreat from the hustle-culture that dominated the pre-pandemic era of giving up everything in pursuit of ambition. Quiet quitting is hard to verify empirically but we all know from experience that it exists. There are many reasons people feel the need to maintain their job at all costs whether it's the health care, the steady paycheck or any of the other benefits that traditional corporate jobs afford. Putting that at risk can be too big a wager to make.

Quiet quitting could be a 'great liberation' in response to the great resignation. People are rejecting overwork and burnout and choosing balance and joy. Employers should take advantage of the quiet quitting movement to encourage a better work–life balance for their staff and communicate to workers that they are valued, leading to greater engagement, productivity, and loyalty.

Quiet quitting is a signal to stop deploring the lack of commitment of workers. This should generate an opportunity to recognize that the proper functioning of organizations depends on what workers do in addition to what is contractually expected of them.

There are still people who are passionate about their jobs, having taken years to build meaningful careers in fields they love, putting in long hours and weathering hard conditions. But in many cases, these dream jobs have become untenable, whether out of toxicity, economic instability or total fatigue. And some workers are asking themselves a big question: is it time to quit the industries they love? Quitting a job suiting one's passion can also lead to a struggle within oneself, vis-à-vis one's sense of identity.

7.7.2 Job hopping

Job hopping to increase salary and skills early in a career appears to be increasingly common. One reason for the prevalence of job hopping is the ongoing erosion of the employer–employee social contract.

For nearly two years, companies have complained that they are caught in an unending cycle of hiring and training workers, only to see them leave in a matter of weeks or months. Constant recruiting and training drains management resources, and new hires often do not stick around long enough for that investment to pay off. Veteran employees are often asked to pick up the slack, leading to burnout.

7.7.3 Great resignation

Post-Covid-19, a wave among academics has many researchers stepping off the tenure-track, mid-career. The pandemic's disruption contributed to a surge in entrepreneurial activity, a key driver of the kind of innovation that could lead to a more productive economy. The dynamics have also spurred many companies to re-evaluate or adapt long-held practices to increase efficiency.

7.7.4 Moonlighting

A moonlighting employee has a 'primary', usually full-time position, and a 'secondary' or part-time position. Safeguarding against job loss is one of the top reasons employees moonlight followed by supplementing incomes. What people want from work has changed forever. It is not just about clocking in hours and going back home. The pandemic has made employees step back and re-evaluate priorities

7.7.5 Working from home

As workers have gotten more exposure to working remotely during Covid-19, more of them want to continue doing it. One reason: they are finding more work–life balance. And a large fraction of those currently working from home, at least part-time, say they want to continue it.

The pandemic made people realize that you don't need to travel for work two hours a day to sit in front of a computer that is connected to the internet anyway.

Some organizations have operated remotely for years with positive results, and their experiences offer lessons for managers and employees working from home for the first time.

Staff should have flexibility to take breaks and run personal errands as long as they complete their work. However, a huge challenge of running a virtual office is building and nurturing a strong organizational culture without in-person interactions.

7.7.6 Corporate sector

Running a virtual office brings unique challenges no matter the context, experts say, but especially during these difficult times, when many companies are battling falling revenue.

The corporate world has changed more in the past two years than in the past 20 years. It is no longer possible to attract people to work at a full-time job in a corporate office because people have realized that the idea of a 'safe and secure' job is just a dream that can collapse at any time. There is no need to work at a specific location in a specific city because we all live in the global village called the internet.

7.7.7 PhD plus skills

Currently, the number of candidates opting for PhDs in AI and machine learning are rising, at the cost of a PhD in physics or mathematics, hoping that one can get hired to work directly on a topic that has generated buzz in industry, just as 'learning to code' was the buzzword a few years ago. Academic career consultants, too, keep advising that the tech industry is about to hire highly educated PhDs.

If you have expertise on a specific skill, you can remotely work for the best companies in the world and command earnings that compete with anyone in the world with the same skill. And the best part is that you can work on a contract basis. People resist becoming freelancers and quitting their day job because freelancing also requires professional relationship skills, sales skills, and the skill of adding more value than what you are getting paid for. These skills are vastly different from the skill of being an employee.

7.7.8 Remote working to entrepreneurship

Freelancing will make you an entrepreneur where the product is you, yourself. This could be the first step in your long journey of building something for yourself.

Already virtual permanence, the transition to virtual workplaces as part of a standard operational structure, is creating broad access to non-local talent pools.

The time of Covid-19 has already brought a much deeper understanding about the meaning of resilience in retaining and attracting new talent. Proactively managing constant change is about keeping pace with the present—by evolving the employee-value proposition through an integrated approach to rewards, workplace culture, and employee well-being.

7.8 Plagiarism versus integrity; fake online reviews; publication charges; patent manipulation; citation multiplication; prolific authors, paper mills; fraudsters

7.8.1 Plagiarism and frauds

To plagiarize means to represent another person's words or ideas as your own. Pretending that someone else's work is your own, i.e., using it without attribution, is plagiarism; the plagiarized item may be a section of text, a figure, a table, or even an idea. And re-using sections of text, figures, tables, or data from your own previously published work without crediting the original publication is also plagiarism, in this case self-plagiarism.

Plagiarism is thus a form of theft (stealing someone else's work) and dishonesty (passing it off as your own). If the copied material is extensive, it may involve copyright theft as well. Plagiarism ultimately weakens the quality of the science and is dangerous. It's a form of corruption

The notion that all researchers must compose their own sentences remains a bedrock principle, but that view might encounter new resistance in a world with essentially limitless access to information and increasingly sophisticated AI algorithms that can reproduce language with eerie accuracy.

Plagiarism-checking software cannot account for academic norms and standardized definitions, nor can it assess whether copied text is truly plagiarism or whether it is central to the conclusions of a paper. It can't catch all instances of matching text, a challenge that will only increase with the use of AI, such as the chatbot ChatGPT, which can swap out words and rewrite text that is fed into it.

The detection of similar or identical strings of text does not always indicate plagiarism; the re-used material may in fact have been credited to its source, or it may represent a standard description of something, such as a common procedure. The editor must always look at the highlighted sections of text and make a decision as to whether or not plagiarism is involved.

Authors can avoid suspicions of plagiarism by always identifying clearly any material that has been taken from another source (whether the author's own work or someone else's). Short quoted extracts should be indicated in quotation marks, while longer ones should be shown as indented paragraphs. In either case, a full citation must be given to the original source; if the extract is at all extensive, copyright permission must also be sought from the publisher as well as the author, since the latter may not always be the copyright holder. Changing a few words or even paraphrasing someone else's words does not alter the situation, since the content is still not original—the original author and source should still be credited.

The internet makes it easier not only to commit plagiarism, but also to detect it. Now that we live in the network and digital age, it is no longer true that seeing is believing. Not so long ago, everyone knew that a photograph doesn't lie. Today, image manipulation is not only possible but common. The fact that research has been published doesn't automatically imply that it's true. Editors of academic journals now have to spend a great deal of time dealing with a variety of forms of author misconduct, in particular plagiarism.

The basic premise of science is to generate facts and theories that can be replicated by anyone interested, with the implication that they are verifiable publicly. The role of a publication in furthering this pursuit is basically to encourage development of such public knowledge. Hence, fraudulent publication is reprehensible in wasting the time of the community, as is a sloppy or inaccurate publication. The root of scientific work is the engagement of widespread independent activity to explore and build upon the published record. If the publication is valid and seminal it will add to the record.

Most labs and their leaders are definitely doing good science that is accurately presented in published articles. Not only must research be held to high standards and scrutiny, but so must the subsequent public relations and marketing. The problem of scientific misconduct though not new, is getting more frequent among scientists. This is at least partly due to the digital technology that enables easy splicing and manipulation of data. Hopefully, running all papers through AI can help to identify if there is fraud.

Historically, academia is among the most honest and ethical professions, which is why it is newsworthy when it is not. Errors should be checked. Errors in published papers when found, make it incumbent on the authors, and the journal editors, to see that the appropriate action is taken. Public trust matters.

Though only a small number of papers take recourse to unreliable methodologies, it is worrisome. A huge number of excellent publications are out every year, reporting accurate results to continue the culture of fostering the development of scientific and engineering advances at the forefront of technology, medicine, and a variety of other fields useful to humanity. Malpractices by authors might undercut public trust in science during a time of increasing skepticism and attacks. But scientists must show that they can handle even rare cases of apparent misconduct and use them as examples to warn the world.

It is true that, if a claimed result is important enough, an inability to replicate it or of subsequent work to conform to it will eventually be noticed. It wouldn't be easy to hide it in the shadows for a long time.

The authors may compose figures by piecing together parts of photos from different experiments. Some supposed manipulation might turn out to be digital artifacts that can occur inadvertently during image processing, a possibility.

Generative AI systems such as LLMs can create new data, including text and images, using models derived from their training data. Researchers can use such algorithms to enhance the resolution of images, for instance. But unless they take great care, they could end up introducing artifacts. Senior scientists should uphold the trust by taking resort to prudent verification.

7.8.2 The peer review process

Before scientific papers are published, they undergo a peer review, a process in which two or three independent scientists judge an article for scientific rigor and correct analysis. Peer review is unpaid and undervalued, and the system is based on a trusting, non-adversarial relationship. Besides, peer review is not set up to detect fraud.

The peer review process, designed to ensure the quality of studies before publication, is based on a foundation of honesty between author and reviewer. Peer review of manuscripts in science was initially started by commercial publishers as a route to legitimatize themselves. The peer review should, in principle, catch cheating or faulty theories or results that don't quite measure up to the claims made. However, it has often failed to catch brazen image manipulation. Concerns about such manipulations, when raised, often fail to gain public attention or to prompt correction of the scientific record.

Peer reviewers stick to an analysis of the proposed scientific facts, and can't be expected to police sophisticated imaging fakery, because that is not their expertise. High-esteem journals should, as a general policy, employ image analysts/fraud detectors to review every accepted manuscript. This would give these journals far better credibility.

The process through which editors ask outside experts to critique a paper's methodology, reporting, and conclusions, is vital to academic publishing. Highly respected academics can receive several peer review requests a day, and may not have time to perform more than a few a month. That means researchers wait for months while their papers languish on an editor's desk. In dire cases, the research may even become outdated before it's published.

Journals are allowing people who have no formal training to peer review. Very few doctoral or research programs actively discuss the importance of peer review, or how to do it well.

To combat delays, the peer review system needs a redesign, and should be incentivized. Some journals do pay reviewers or waive their article fees or subscription costs. For example, submitting three or more reviews per year for a Nature Research journal entitles reviewers to a free, online subscription to the journal they choose from Nature Research's offerings. The Springer Nature reviewers often appreciate recognition in the form of subscriptions, and having their names included in the published papers.

7.8.3 Citation manipulation

Publishers should have a clear policy on citation manipulation. Editors must not attempt to boost the ranking of their journals by artificially inflating any metrics, or coercing the authors to cite certain publications, in return for facilitating the acceptance of their papers. Thus, the researchers who agree to manipulate citations are more likely to get their papers published.

7.8.4 Words of caution to STEM PhD candidates

PhD students need appropriate education and mentoring so as to avoid making naïve mistakes.

Doctoral and postdoctoral researchers being often subject to the intense pressure of the need to publish or perish is a common thing in academic science. Getting a paper with your name on it in Nature, Science or Cell, the high-profile journals can make or break young careers. They are also useful to stand out in a field with limited lab positions and professorship openings. Senior researchers often take credit for the work and ideas of their postdocs, but brush off responsibility should errors or mistakes arise. Falsification, involving playing with research standards is luckily rare and not a norm, yet.

There's undoubtedly too much pressure to just produce anything publishable. But maintaining the impression of constant excellence by an institution requires regular production of high-quality science.

Ironically, it is the average authors, and not leaders in a field, who find their research paper more closely scrutinized when they submit it for publication. A manuscript sent to high impact-factor journals have a higher likelihood of being accepted for review if submitted from an Ivy League institution.

Scientific ethics are more important than salary or professional advancement. You should possibly walk away from a situation where you feel pressured to fabricate data, more so when you are working on something which could make a huge impact on society or impacting human health. In an industry setting, too, when a young scientist is put on a project, and he finds inconsistencies, it takes integrity and conviction for him to point to falsehoods in data analysis.

Trust, honesty, and truth are precious things. Fraud and deceit are the tools used by inferior individuals to pull down the idea of integrity.

7.8.5 Plagiarism, publication bias

Arguments can be made that every professor should have their dissertation/thesis checked for plagiarism. But hidden in this suggestion may be a plot to tar the school as a place where plagiarists teach. Plagiarism being used as a weapon will only result in the public not taking plagiarism more seriously, but less. As individuals become inundated with an ever-increasing number of questionable plagiarism stories, the serious cases may end up getting lost in the noise.

Dr Claudine Gay, Harvard's first black president and the second woman to lead the university, ended her turbulent tenure that began in July 2023. It is the shortest stint in office of any Harvard president since its founding in 1636. Prior to her resignation, when the president of Harvard, faced allegations of plagiarism, the university suggested that she was guilty only of 'inadequate citation' and 'duplicative language without appropriate attribution' and said that no evidence of 'intentional deception or recklessness' had been found. Anyone who believes they should be exonerated on charges of plagiarism because they had no ill intent can marshal the examples of Ivy League academics in their defense [9]. Some scholars were skeptical of the plagiarism allegations, saying Gay had not taken credit for others' original ideas or data, but just had minor echoes of jargon and routine language from political science. Others, though, said the growing number of allegations was troubling, and asked whether she had been held to a less-strict standard than the university's students would be. However, some academics felt that she was targeted, criticized, and essentially driven from the job largely because of her race. Some believed that, in accordance with the precepts of DEI (diversity, equity, and inclusion), Gay had been appointed as Harvard president more for her skin color than for her professional qualifications.

Subsequently, Claudine Gay came under fire for seeming to equivocate before Congress when asked to discuss Harvard's antisemitism policies.

The Claudine Gay debacle at Harvard has raised some fundamental questions about academia in general. She was president of the university, traditionally seen as the pinnacle of American academia, and it became easy to chalk up the Claudine Gay situation to affirmative action.

7.8.6 'Prolific' authors, article publication charges, predatory journals, paper mills, retractions

Of late, there has been a jump in number of 'extremely productive' authors. A finger of doubt is raised whether all their manuscripts were produced through honest labor, or questionable research practices were used. It is suspected that some of such researchers pay for authorship, and list their names on papers they had bought.

The combination of 'publish or perish' culture and incentives offered 'supposedly' by some universities create fertile grounds for shady authors/publishers to flourish. During the pandemic, paper mills flourished which sold fake papers to researchers.

Machine learning (ML) and AI are powerful statistical tools that can pick out patterns in data that are often invisible to human researchers. At the same time, it is likely that ill-informed use of AI software may be driving a deluge of papers with claims that cannot be replicated, or are practically useless. AI provides a tool that allows researchers to 'play' with the data and parameters until the results are aligned with the expectations. This flexibility and tunability of AI, and the lack of rigor in developing these models, provide way too much latitude.

7.8.6.1 Hyper-authorship

The term 'hyper-authorship' is used to describe papers which have 100 or more co-authors. With the rise of large international and multi-institutional scientific collaborations, such as the ATLAS consortium behind the discovery of the Higgs boson, papers with hundreds of authors are becoming common. There may be legitimate reasons behind it, but it is raising questions, and concerns, about the nature of authorship and the impact that hyper-authorship has on the metrics of scientific achievement.

Hyper-authorship is also the result of scientists in some fields seeking answers that require not just large-scale collaboration, but also huge resources and equipment. This is particularly evident in high-energy physics, where the cost of equipment such as particle accelerators can run into billions of dollars.

The increasing length of author lists, however, does lead to the question of what level of contribution entitles researchers to be included. How do you actually quantify what authorship means in terms of what contribution someone made, and then how that is used as a currency in terms of what it means for hiring or promotions?

7.8.6.2 Paper mills

An unrealistic publication record is often needed to obtain the best jobs, and those who publish in top journals also get big cash bonuses. That has led to thriving paper mills. Authorships are up for sale and editors of lesser-known journals offer to publish papers for payment. And for the right price, there are ghost writers who will write an entire research paper for a 'client'. Ghost writers act as fixers for academics caught in the endless cycle of publishing and promotions, and PhD scholars desperate for their degrees.

Paper mills are scientific shops that sell everything from authorships to entire manuscripts to researchers who need to publish lest they perish. Paper mills charge a fee for fabricating data about basic science. Such sham papers deserve to be retracted. Teams of people who write, edit, and proof ambiguous articles make it impossible to determine who's getting paid and for what.

In recent years, publishers have stepped up efforts to tackle the output from paper mills. Not only these are of poor quality, the articles from paper mills often contain made-up data and nonsensical text. Some paper mills take pride in claiming to have brokered tens of thousands of authorships—including in journals that are indexed in respected databases.

Highly skilled researchers are generally busy in their own work and have no reason to spend their time trying to replicate other peoples' research, for checking if any malpractices have been incorporated. An alternative is to make a replication study of an important paper a prerequisite for a PhD degree in an empirical field. Such a requirement would allow students to see first-hand how research is done and would also generate thousands of replication tests.

'Contract cheating' is a continuing problem for higher education, as a global industry is targeting students with unsolicited offers to write bespoke essays or to complete assignments to tight deadlines. These are often designed to get past plagiarism-detection software.

7.8.6.3 Predatory journals

Predatory journals are a scourge of science. They collect publication fees and publish articles without adequate (or sometimes any) peer review, ultimately wasting researchers' time and money and undermining public trust in science.

Predatory publishers hook uninformed researchers to scientific publishing that is fast and obstacle-free. Even when some authors try to withdraw their papers before publication, having noticed the fraud, the predatory journals publish them anyway.

Predatory journals prey on the gaps in institutional policies. Only through vigilance can institutions support and protect their researchers from harassment, fraud and the waste of papers that, in the hands of legitimate publishers, could have contributed to the world.

7.8.6.4 Retractions

The number of retractions issued for research articles in 2023 passed 10 000, a new record. An analysis by Nature has found that among large research-producing nations, Saudi Arabia, Pakistan, Russia, and China have the highest retraction rates over the past two decades [10]. Although the majority of retractions are due to misconduct, this is not always the case, some are from authors who discovered honest errors in their work.

Often, problems with data—tables, statistical tests, charts, and photos—are not caught until after publication. A paper should be retracted if critics can demonstrate scientific misconduct such as photoshopping or faked data. After retraction, it will still be available to read or download but will be marked as untrustworthy. Unfortunately, many scientific journals and academic institutions are slow to respond to evidence of image manipulation—if they take action at all.

A cottage industry of checking research papers had already sprung up in the last two decades, including Retraction Watch, a blog dedicated to unmasking research based on bad data. In a move to protect researchers from scams, Retraction Watch, working for issues of scientific integrity, created a database of fake journal sites [11].

7.8.6.5 Scientific integrity

In the world of academic research, the only goal should be to obtain truth. Researchers gain a lot of prestige with each successful publication. For honest research findings, increased funding and citations boost their reputation and status—accolades which are well-deserved.

Scientific research is a laborious and intensely challenging process. Therefore, rewarding notable scientific progress through publication and career advancement is crucial to advance knowledge.

Science needs all the credibility it can muster. Therefore, ensuring scientific integrity and earning public trust should be the highest priority. This self-reflection in the scientific research community is important. To address research misconduct, it must first be brought into the light and examined in the open. The underlying reasons scientists might feel tempted to cheat must be thoroughly understood.

Scientific integrity is the religion of scientists. The falsification of data is appalling. Even professors who are in the lab full time overseeing everything can be fooled by a con man altering the data. Training in scientific integrity must be strengthened, both by teaching young scientists to maintain credible research practices and by training all researchers to identify and report scientific misconduct.

The cracks in scientific honesty and rigor that seem to be on the rise are caused by several forces at research universities. For many years now 'publish or perish' has actually taken back seat to 'bring in as many dollars' as possible to your university. Of course, publications in Cell, Science and Nature-like journals are the honey that impress the primary sources of those dollars. Do the public understand that for all the funds a scientist brings to his institution, his university receives a part of it as overheads? Unfortunately, scientists don't get rewarded for a high level of integrity resulting in careful and honest research, they get rewarded for blockbuster papers.

Scientific research is far too important to be subject to the push and pull of egos and ambition without safeguards.

The fundamental bedrocks of research are honesty and trust. The senior author has multiple time-consuming responsibilities: overseeing the lab, administrative meetings, writing grant applications, teaching, etc, and hence cannot micromanage every detail of methodology, data collection, and statistics, etc. This emphasizes the critical importance of the reviewer at the journals to review in detail each component of the article and 'not' article with minimal comments, questions or concerns.

How many authors of current publications are really and completely involved in the whole experiment that led to the data reported? Some of them just made a measurement, maybe without being aware about the experimental hypothesis and the discussion of the results. Until some years ago we used to just acknowledge such collaborators.

Starting in the mid-20th century, companies began distorting and manipulating science to favor specific commercial interests. When strong evidence that smoking caused lung cancer emerged in the 1950s, the tobacco industry began a campaign to obscure this fact. The entire industry of evidence-based (and academic) medicine is now suspect due to the malfeasance of certain pharma players.

7.8.6.6 Open science

Open science is a broad term that refers to the movement of making the entire research life cycle freely available to everyone, from citizens and students to research professionals. This includes sharing research plans, protocols, materials, data, and papers through open-access platforms. The practice of open science is on an upswing.

Article publication charges (APC) have become an integral part of the open access—the system in which papers are freely available for all to read after publication. The non-profit open-access publisher PLOS, based in San Francisco, California, is experimenting with some non-APC ways of publishing [12]. US President Joe Biden has announced a commitment to make any federally funded research results freely available to the public, starting in 2026.

The Center for Open Science, a non-profit organization in Charlottesville, Virginia, aims to achieve open science by following a strategy consisting of five steps: making open-science practices possible, easy, normal, rewardable, and eventually a requirement.

7.9 Positivity in job search; networking benefits; workplace visibility; avoiding getting scooped; brain breaks; follow your passion; learning AI and ML skills

7.9.1 Working to be employable

A PhD scholar goes through a dissertation, does postdoc work, and perhaps a series of visiting professorships. That scholar, if they land a tenure-track position, might be in their 30s before starting it.

It is important to focus on employability. What makes the biggest difference in terms of being a strong professional asset is work ethic, passion, and interpersonal skills. If you can learn to be flexible and think critically, you'll be just fine, no matter which school you came from. Those are the skills that will see you through the 30–40 years of employment ahead of you. To continue to remain employable, it would be rewarding to develop interdisciplinary strengths during PhD research. Many of the complex problems of our world, say climate change to housing affordability—require multiple disciplines to be involved in finding solutions.

7.9.2 Job search with a positive mindset

After you complete your PhD, if you are hired as an assistant professor, at a leading research institution like those from the Ivy league, it is not likely that you'll get tenure down the line. For tenure at a top school, you need to stand among the foremost leaders in your field, by establishing that reputation in the space of say, six years. The chances are you'll stay awhile, publish, then jump to another job somewhere else, somewhere that *will* tenure you. On the other hand, a good job is essential for contentment, a job that gives meaning to life and a sense that one has value. A bad job, meaningless and repetitive, is depressing and eats into one's feeling of self-value.

Tackling the challenge of finding the right job begins with developing an awareness of existing obstacles. Scientists can then use time-tested strategies to stand out from other candidates to land job offers. Competition for jobs is mounting. So, how should STEM PhDs make themselves particularly employable and stand out in an increasingly competitive market? Career specialists have the following points of advice:

(a) One should keep in mind that an average job search currently takes around three to six months, so people should prepare themselves for it. In the meantime, connect with supportive people to stay confident, and learn to suppress any feelings of desperation, because that can be damaging. Desperation should not lead you as a candidate to prematurely lower your standards for the type of job or company culture you'll accept.

(b) Engage yourself in positive activities, say meditation, that build your confidence and capabilities. Volunteering to help others improves your mental and physical health, and thereby improves your chances of finding employment. During difficult times, get in touch with childhood or college friends, who can make a heavy situation lighter and humorous.

(c) Employers usually know that the person they hire won't have everything they're asking for. Hence, just because you don't meet all the criteria listed in a job advertisement, doesn't mean that you needn't apply. Employers often list the desired skills and attributes. Candidates are likely to be considered for a job even if they have about 75% of the items on an employer's wish list. Keywords listing the skills and competencies in the advert usually mean they're of higher importance.

(d) Eliminate negative views that might be holding you back. Rest assured, you are not the only one experiencing self-doubt or apprehensions. Thoughts like, 'I can't do it', 'I don't have the energy', 'I'm not creative', 'I'm not really sure what I want', 'people don't like me', 'I'm not good enough', should be buried. Banish any damaging belief which could become your main obstacle.
(e) Scientists often have a cringe factor associated with self-promotion. They think their work will speak for itself. It is not true. A scientist, too, will have to promote themselves and their strengths for the job that they applied for. A job search is, in fact, quite like scientific research, where one needs to pose questions and gather information.
(f) Look at it from the angle of the company that wants to employ you. If you demonstrate your uniqueness and how that aligns with the employer's needs and company culture, it will wipe away your competition.
(g) Job seekers should identify their unique traits. Remind yourself of a past situation and identify the tasks that you had to do then, the actions you took and the results. Being in a position when you can do some of the things with ease, can relax you and keep you on the right track.
(h) Go for 'networking', or at least build a 'relationship' in the context of a job hunt. A personal connection might get you around the application process. Someone who knows the right people might put your materials in front of the hiring manager. Create genuine connection first. Don't send your CV on LinkedIn without interacting to form a relationship first. Why should a person look at your CV when they don't even know you ?
(i) Avoid submitting your application in haste without really going through the job description and adapting your résumé to match a job in industry. Your résumé should not be a list of all things done by you previously (as is common while writing a CV while applying for an academic job in a university). Make it easy for the employers by conveying three points: (i) the skills you possess, (ii) the evidence that you're successful at those skills, and (iii) why you're successful.
(j) Lastly, don't forget to let a trusted friend, a mentor, or a career professional evaluate if the application you wish to send is clear and concise.

7.9.3 Avoid getting scooped

It takes time to write grant applications, get decisions on whether they have been approved, publish manuscripts and prepare talks, alongside other commitments. It's a perfect opportunity for someone with more time, power, and bandwidth to take your idea and run away with it. How to guard against it:
(a) Get a citable paper published based on your work (say in an Open Access journal) to reach out to a broad research community sooner, to seek priority and ensure that others give due credit to you. You are staking a claim on your research ideas, which you can elaborate later.

(b) In a profession that is by nature competitive, where academic jobs are few and hard to get, there will always be people who cheat and investigators that fail to look behind the curtain. Therefore, try getting first authorship on big papers from a prestigious lab, which would get you more attention and help you land a dream job.

When not getting the job, reflect on what you may be doing wrong in the interview. That should make you do better in future both personally as well as professionally.

7.9.4 Keep relaxed, have brain breaks

Out of a 45-min block for classes, picking out a one minute physically active brain break makes a huge difference for learning. A cognitive downtime helps to process, organize, and integrate the new information just learned, in your mind, and helps to foil 'cognitive fatigue'. That optimizes storage and recall of the new information, when needed.

7.10 Keeping the edge in employability

7.10.1 STEM has an edge for getting jobs

New advancements and discoveries in science and technology are rapidly changing the world of work and increasing the demand for technically skilled employees. The number of STEM workers is also increasing, accordingly. STEM workforce fuels innovation, and therefore the demand for STEM workers goes up steadily. Today, a substantial portion of the graduate workforce is employed in STEM occupations.

A PhD expertize gives you the valuable skills of critical thinking and independent analysis. You can evaluate, analyze, and create solutions to a problem. That is often as valuable as specific subject knowledge.

A STEM PhD degree can open several doors to pursue a number of career paths. Graduate school teaches a doctoral student many skills essential for any career path, such as communication, critical thinking, problem solving, and data analysis. Each of these skills will in all probability be useful in a non-research career path, too. Therefore, continue to build these skills while in your current position, as they will help in your success later on.

STEM workers have higher median earnings and lower rates of unemployment compared with non-STEM workers. The reason for this is a robust STEM knowledge-base and expertise, arising from contributions in innovation and creativity, along with technical skills in STEM.

7.10.2 Continue to learn fresh skills

AI and ML are having a major impact on the tools and platforms we use regardless of whether we're designing products, or targeting image processing. AI has provided engineers with powerful tools to automate tasks, optimize designs, and solve complex problems, which have led to increased efficiency and innovation.

The integration of AI in the engineering workplace represents a transformative approach.

The emergence of AI in the past few years has created fascinating possibilities to apply it to manufacturing issues. STEM PhDs should realize that AI also has its drawbacks, as it raises concerns about privacy, security, and biases within the algorithms themselves. Ethics is also a serious concern, certainly when it pertains to intellectual and proprietary technologies and those who create them.

7.10.3 Keep workplace visibility

Workplace visibility is vital to getting your name registered in the minds of decision-makers who direct career-shaping projects, and eventually, help in landing a promotion, too. Workplace visibility means that your work is noticed, acknowledged, and valued. To be more visible at work:

   (i) Volunteer for learning opportunities that will expand your skills. Seek learning opportunities that will expand your skills. It could be cross-departmental job rotation or a short-time work on a sensitive project.

  (ii) Demonstrate that you have a skillset that is valuable for their work, so that your work will get the attention it deserves.

 (iii) Do quality work to earn the reputation as a reliable, trustworthy, valued, and needed member of the team—all of which leads to more visibility.

It takes more than hard work to build credibility and visibility. Workplace visibility gets you more influence—a must-have for any aspiring leader. Link up with other visible superstars at work, to connect and build relationships with those you admire with the intention of understanding how they earned their visibility. Work alongside them, with them, or for them. Your contributions will become more visible then.

A number of STEM PhDs don't get quickly hired because they didn't have 'relevant experience'. And they can't get relevant experience because they couldn't get hired. To get over this 'circular' argument, recruiting companies should offer to train more candidates.

7.10.4 The key areas to focus on for STEM jobs

Currently, the biggest drivers of job growth are Big-Data analytics; climate change and environmental management technologies; and, encryption and cybersecurity. Agriculture technologies, digital platforms and apps, e-commerce and digital trade, and AI. These are all expected to result in significant market disruption, with job displacement in their organizations, offset by job growth elsewhere, to result in a net positive. The human–machine frontier has shifted, and a third of all business-related tasks are performed by machines, and the remaining two-thirds performed by humans.

The fastest growing roles relative to their size today are driven by technology, digitalization, and sustainability. The majority of the fastest growing roles are technology-related roles. AI and ML specialists top the list of fast-growing jobs, followed by sustainability specialists, business intelligence analysts, and information security analysts. Renewable energy engineers, and solar energy installation and system engineers are relatively fast-growing roles, as economies shift towards renewable energy.

Large-scale job growth is also expected in education, agriculture, and digital commerce and trade. Jobs in the education industry are expected to grow by about 10%, leading to a huge number of additional jobs for vocational education teachers and university and higher education teachers.

Generative AI has received particular attention recently, with claims that 19% of the workforce could have over 50% of their tasks automated by AI. The employment for e-commerce specialists, digital transformation specialists, and digital marketing and strategy specialists is expected to increase by over 20%, leading to an availability of thousands of jobs. Demand for AI and ML specialists is expected to grow by 40%, which means more jobs, as the usage of AI and ML would drive a continued industry transformation [13].

7.11 A job in industry for STEM PhDs, or even a postdoc in industry

PhD candidates often hesitate to apply for a job in industry, considering that they may lack the skills necessary for transition from academia to industry. Fresh PhDs can exploit their extensive expertise and know-how by securing employment in industry. Considering that most PhDs will never get a tenure-track job in academia, such as a research university, they should come out of the mindset to try only for academic jobs. It is worthwhile to take up the challenge to try for diverse career paths, which are aplenty in industry.

7.11.1 Engineering remains a viable and important career path in industry

Staying current by learning new and emerging technologies is a top priority of professional engineers. There is always a need for finding qualified engineering candidates who are keen to master new, useful skills. For instance, engineers with mechanical design as a specialty are always short in supply versus their demand, hence are most sought-after during recruitment of engineers.

The use of engineering videos as a source of information continues to grow, because a majority of aspiring engineers use videos to further their engineering education. That is followed by engineering/technology publications. Trade publications such as Machine Design are a trusted resource list, whether as hard copy publications or soft literature at the publication website. The last option is seminars.

In some ways the migration to a more flexible workplace environment can benefit a working professional, but it depends how the company involved establishes the rules and relationships under which their employees work. When asked about job satisfaction, most of the machine design respondents describe themselves as very satisfied with their jobs.

One of the biggest impacts of the pandemic was how it changed the way people work. Before the Covid-19, an engineer was expected to attend an office every day, but there are now a lot of them working from a home office, as well.

In the following sections, we attempt to focus on EVs, fuel cells and AI-supported self-driven cars, as a case study for jobs opportunities for STEM PhDs.

7.11.2 EV profession as a case study for job growth

The feeling of never again needing to stop your car for gas is life changing. A popular EV like KIA EV6 is a convenience for all the reasons and does not disappoint. Home charging it costs just 4–5 cents a mile. However, it would cost a fortune transforming the generating grid across the world to renewables and upgrade the transmission grid. When you charge your car's battery, you are probably directly linking to oil or gas. The manufacture of electric vehicles is the third highest release of CO_2 in all manufacturing.

The battery costs have come down a lot in recent years, and the essential controls for operating the vehicles are known well, too. EV technologies are getting increasingly commercialized and the charging infrastructure is adequate for EV owners, though yet only in the large cities. Despite mastering the essential technologies for building electric vehicles, it is still a challenge for EV start-ups to master mass manufacturing techniques, and supply chain issues.

Besides, the industry has not yet solved the issues of battery degradation and the financial impacts of battery replacement.

Manufacturers advertise 330-mile range and 90-min charging. That means over 20% of a driver's and truck's time is spent charging, time spent not moving or meeting schedules, while the debt and salary meters are still running.

7.11.2.1 H_2 fuel cells

H_2 is cryogenic and takes a lot of energy to produce as a compressed gas or liquid. Using today's technology, liquefaction consumes more than 30% of the energy content of the hydrogen and is expensive. It's also extremely flammable or explosive and as the temperature is so low it can condense oxygen from the air into a liquid which easily ignites and will burn with almost anything including metals.

It still takes more energy to produce hydrogen than hydrogen produces when it is burned. There are circumstances in which hydrogen fuel cells are better suited than fossil fuel energy sources and worth the increased price, but so far fossil fuels have always gone into producing the hydrogen, whether it is derived from methane or electrolysis.

Looking at the current scene of H_2 fuel cell technology, it is possibly a good idea for a STEM PhD to take up a R & D position in a budding H_2 fuel cell company of start-up.

7.11.2.2 AI-supported self-driven cars

Hundreds of thousands of deaths every year on roads around the world suggests that self-driving cars is not yet a solution. Tesla is working on self-driving based on AI's ability to think about situations it will encounter on the road and react in ways a human driver trained by billions of miles of the best drivers, based on the data collected from the millions of Tesla vehicles currently on the road, would react.

The problem is that the public expects such technology should be in its perfection rather than just better than a human driver.

7.12 Concluding remarks

The structure of doctoral education has barely altered in past few decades, and the goal of graduate school has always been structured to prepare people to become research scholars. It only grooms students for high-prestige positions at research universities.

While selecting their major subjects, young college students are often advised to follow their passion, even though they don't have much idea about which subjects they are passionate about. Often times, men pick up science and women pick up arts largely because that is what society expects from them. But that leads to gender gaps, particularly in engineering subjects and in computer science, where women are already underrepresented. Students need to be exposed to a variety of options before they enter college, so that they get a chance to gauge the potential of STEM and make choices which are not only creative but also offer them financially rewarding careers.

In academia, the tenure system of advancement and the standing of universities depends on the integrity of the review system and quality of the research. Without that, science and engineering cannot advance.

Quality of output is the only thing that is important for an engineer, including STEM-trained, in an industry job, including STEM-oriented industry jobs. Therefore, education with a focus on learning new relevant skills is the key. Useful skills for employment are critical thinking, collaborative problem solving, and, yes, understanding and learning to how to address systemic forms of discrimination, especially related to race and gender. Your ability to communicate at all levels also plays a major part in how you're going to advance your career in industry. Maintaining good communication with co-workers across levels, from lab technicians to upper management, and across disciplines and departments, helps.

A PhD from a highly rated university will be of benefit, while seeking a job in academe, because the CV of such a candidate will reflect the pedigree.

A 3-year assistant professor position looks fancier than a postdoc but the chances of progressing to a permanent job are no greater, vis-à-vis. After three years, the Assistant Professor will be jobless again, existing in a state of impermanence, suffering emotional costs, which could be accompanied by the long wait for stability before starting a family, etc.

Career dissatisfaction grows as one advances as a postdoc from second to third and to fourth postdoc position. After years of postdocs, the candidates enter their thirties and become more negative about job prospects, job security, and work–life balance than those under 30, and are likely to undergo mental-health challenges.

Funders from high-income countries should discourage funding of research under which well-qualified researchers work in lower-income settings, where the PIs fail to involve local researchers in all stages of the research.

All stakeholders, from researchers, institutions, funders, and policy-makers ought to establish long-term relationships with collaborators that extend beyond the life of a single project, thereby helping to remove obstacles between postdocs and regular jobs [14]. Policy-makers must commit to green skills, and prepare the workforce, accordingly. The green skills revolution has the potential to transform our working lives in the same way the rise of the internet and digital connectivity did. Business leaders must invest in upskilling current and future green talent. The global workforce has to build green skills to go for the best jobs.

References

[1] ECU 2023 *East Carolina University, University Affairs Committee, Agenda Report* p 62 https://board-of-trustees.ecu.edu/wp-content/pv-uploads/sites/121/2023/04/University-Affairs-Committee-2.pdf

[2] Griffith E 2023 From unicorns to zombies: tech start-ups run out of time and money *New York Times* https://nytimes.com/2023/12/07/technology/tech-startups-collapse.html

[3] Brown E 2016 The fight for accessibility *Nature* **532** 137–9

[4] Spoon K *et al* 2023 Gender and retention patterns among U.S. faculty *Sci. Adv.* **9** eadi2205

[5] *'Loss for the nation': Japan rushes to erase stigma for women in STEM fields, SCMP* (2023) https://scmp.com/news/asia/east-asia/article/3227471/loss-nation-japan-rushes-erase-stigma-women-stem-fields

[6] *Snapshots of disparity in STEM, Advancing Women in STEM Sratefy, Department of Industry Science and Resources, Australia* https://industry.gov.au/publications/advancing-women-stem-strategy/snapshot-disparity-stem

[7] Durrani J 2023 Small pilot scheme will support women returning to stem careers in UK *Chemistry World* https://chemistryworld.com/news/small-pilot-scheme-will-support-women-returning-to-stem-careers-in-uk/4017028.article

[8] NSF Special Report 2023 Diversity and STEM—women, minorities, and persons with disabilities *National Center for Science and Engineering Statistics, Directorate for Social, Behavioural and Economic Science* pp 23–315 https://ncses.nsf.gov/pubs/nsf23315/report

[9] Hartocollis A 2024 The next battle in higher Ed May strike at its soul: scholarship *New York Times* https://nytimes.com/2024/01/14/us/plagiarism-harvard-claudine-gay-neri-oxman.html

[10] Van Noorden R 2023 More than 10,000 research papers were retracted in 2023—a new record *Nature* **624** 479–81

[11] https://retractionwatch.com/

[12] Sanderson K 2023 Who should pay for open-access publishing? APC alternatives emerge *Nature* **623** 472–3

[13] The Future of Jobs Report 2023 *World Economic Forum* https://weforum.org/reports/the-future-of-jobs-report-2023/

[14] Clancy M *et al* 2023 To speed scientific progress, understand how science policy works *Nature* **620** 724–6

Chapter 8

Growing to be a leader and staying on top

We shall discuss how to use every opportunity to build self-confidence through productivity, and translate it into positive energy for self and for colleagues at work. In the first half (section 8.1) of this chapter, we discuss the routes taken generally by a person after securing a PhD degree in STEM, and perhaps a stint of postdoc work, for being chosen to lead a group at a University, or Head a Section in industry, or a government institution. In the second half (section 8.2), we discuss hints and methodologies adopted to consolidate one's leadership role, and to grow further in one's organization, and to stay on top of it, to help in steering your organization to greater heights, despite challenges and crises faced on the way.

8.1 Working towards a leadership role in a STEM job

8.1.1 Getting noticed, gaining visibility

To move ahead in their careers, postdocs need to develop visibility of their scientific achievements, for which the first requirement is to keep publishing research papers and obtain citations on these papers.

The fastest way to success isn't getting a fancy degree. It's doing and learning more than your peers, and getting noticed in the workplace, cheerfully and relentlessly. Volunteer for every little task. Whenever someone is looking for help, offer it. The more people associate you with getting things done, the more top-of-mind you'll be for the next big task.

Learn a lot. Study what's happening and why. Ask questions. Watch and take notes on people you admire.

A step in the ladder of leadership is often to be a member of a student committee or a conference organizing committee. Both these can provide an opportunity to demonstrate and hone one's management skills, and get noticed by supervisors or employers.

As a candidate, build research and other collaborative partnerships with faculty members in your department so they are likely to speak up on your behalf when your promotion file is next discussed. Be your own best advocate. Explain what you do and the impact of your research, over and above your publications.

8.1.2 Leadership through multidisciplinary talents for collaborations in a scientific career

Every day spent in the lab as a postdoc is a learning experience, a training which presents the baby-steps of the path to leadership—learning to lead. Postdocs learn to interact with scientists from diverse disciplines—physics, chemistry, political science, and sociology. This way they develop skills to communicate across disciplines, a key factor needed for leadership. Leadership skills are not only useful but essential for a postdoc to rise in their career.

Postdocs who don't develop leadership skills get left out even if they have done well in their own subject, because having a multidisciplinary bent of mind is essential to be able to interact and eventually collaborate across different disciplines of STEM, or even with experts working outside STEM subjects to grow in one's own career and move up the ladder in it. Remaining a specialist in one's own subject will bring a postdoc to being in a 'single subject cocoon', and remain largely unproductive in their career. The future of an active and expanding career is in collaborating with experts beyond your own subject speciality.

8.1.3 Developing skills for leadership

One can start working on one's leadership skills, at any stage of a scientific career, to be ready when a chance comes up to lead one's own lab, and maybe more. Perhaps the most appropriate stage of learning to lead is as a postdoc.

Leadership lessons abound at every corner. To be a leader, reflect on what you observe. A leader will spot patterns in problems. Remaining on the lookout for potential lessons is key to making the leap to leadership and flourishing once you get there. The following personal qualities will help to grow leadership skills:

(A) Before interacting with others in the lab first know and assess your own capabilities. Do you have the potential to help people around you, and gain popularity with them?

(B) Learn to exercise emotional intelligence. Put yourself in the place of a workmate who is not able to perform well. Try to learn about the difficulties that colleague is facing. Try to be of help. Sometimes, an offer of help cheers up a person, and he gets back to their full potential to working.

(C) If you wish to see yourself as a leader, then stay positive in order to keep your colleagues motivated, whether they are MSc/PhD students in STEM, or technicians in the lab. Exuding optimism gives a boost to the productivity of people around you.

(D) Don't be rigid. Be flexible during discussions within your own group at your workplace or while presenting a project report to your seniors, or to funding committees, and venture capitalists. At no stage should you give the impression that you have made up your mind and are not amenable to any further suggestions.
(E) Listen to colleagues and employees. Employees want to be heard. Listening has become a crucial skill for leaders. People like to follow leaders who are listening more than pronouncing.
(F) Be decisive. Employees expect leaders to take decisions, particularly during a crisis, when people value directional guidance—not micromanagement.
(G) Explain the rationale. As a leader, you must explain your thinking about crucial matters. People accept to work on a directive when they understand the rationale and how it fits the bigger picture.
(H) Clarity of vision. Be clear about your vision and strategy while sharing your plans with staff and colleagues. Try to connect the vision to what people are doing every day.
(I) Convince to lead. Master the art of persuasion to convince members of your team to agree to a commitment, or to accept a decision or a course of action. How? Reach out to the right people in the right sequence, keeping in mind what you want from each person. Key points are: grab their attention; give them credit; practice fairness.

8.1.4 Networking for leadership

To become a leader, a scientist must be good at networking, viz. developing strong connections with colleagues around them, with a selfless approach, in order to first develop a visibility so that others in the lab, both senior and junior know that you exist, and are ever-available to help.

Networking is when you converse with others who have something to offer, and vice-versa. MBA students have long prized off-campus relationship building as a key part of a degree that can cost them as much as $200 000 at certain schools. Those networking opportunities might happen on ski trips, at black-tie galas, or informal summertime group gatherings, such as Yacht Week to get to know one another before business school begins. Young professionals have met fellow MBAs and entrepreneurs, during a Yacht Week as they sleep in tight quarters or above deck on the fleet of boats which move between islands. In such a setting, people are a lot more open to connecting and having conversations.

8.1.5 Collaborations

Finding great collaborators, and being able to work with them productively, is one of the most important predictors of success. For that one needs to learn about prospective collaborators' working styles and ambitions—and to get a sense of how well you 'click' on a social level. Most of all, a collaborator must contribute.

A collaborator must also be looking at and trying to evaluate you, as a collaborator, likewise. So, the partners in a collaboration needs to value the contributions of either side, and be complementary, and not disproportionate in inputs made. And also not be a burden on the other side [1].

8.1.6 When an offer comes, grab it

Succeeding as a top leader has little to do with your title and everything to do with your mindset. If the dean called tomorrow and asked you to become your department's new chair, would you be prepared, willing, and able? YES, anyone with the skills and inclination can step up and lead. What to do if you're suddenly offered a management position, what factors to consider before you say yes? Sudden leadership transitions have always happened in higher education.

Most people who become administrators overnight were not completely shocked to receive the offer. In fact, the choice to call upon them was already a possibility. Their capabilities such as their interest in management, and their willingness to serve were shaping up. They chaired search committees well, they sat on a university task force successfully, they helped to design a new academic program. In short, they had done the work and indicated that they liked it. They had also set up conditions conducive for opportunity to come knocking.

On two occasions in my own case, I was asked on the phone, if I would be ready to take up a higher post giving me control of larger scientific, or administrative activities. I agreed immediately, knowing well that my seniors must have valued my performance and examined my potential before making this offer, so it was.

Build a management network. Keep prepared for that moment when the offer comes, by assembling a stable of trusted colleagues whom you consult for advice when it does. Your pool of allies should include both people like you and mentors who are already campus leaders.

Pay attention to what the people you care most about are willing to give up, in the near or distant future, for your career.

8.1.7 Pay back with gratitude

We receive scholarly gifts from those 'above' us in the profession, whether they are of higher rank, have more seniority or greater professional stature, or are associated with a more prestigious institution. The gift could be, for example, a valuable letter of recommendation. An ethical obligation to give back is combined with a structural inability to repay directly those you owe, since there's no way you can repay them directly, simply because anything as your 'gift' is no good in their realm. How, then, do we even begin to pay back these scholarly gifts? The answer, in short, is that we turn around and pay them down the line: pay them to younger or less well-situated scholars we are in a position to help.

Our giving back could be through offering our scholarly capital to those who have less, with the understanding that they'll at some point turn around and make that same investment in others. Or a generosity that is more or less obligatory, implicitly a condition of employment (serving on dissertation committees, writing letters of recommendation when asked).

8.1.8 A position at the top brings administrative workload

At a certain stage of our careers, administrative work is no longer something to dread or to apologize for. Maybe, it often calls us away from our academic regular work. Academic administration can be a calling; that work can be incredibly rewarding instead of draining or distracting; though it requires training and accomplishment as a scholar to qualify for such an appointment.

One never gets fully compensated for the additional administrative work. Some institutions offer department heads a small additional stipend; some reduce the chair's teaching load. Some do both. I once overheard two very senior academics discussing the perils of the administrative work that came their way when they became Department Heads. One of them remarked: 'And the pittance of salary increase I got in return is not even enough to buy aspirin'.

People often talk disparagingly of administrative academic posts, stating that those who can (teach, research, write), do; those who can't, or can no longer, chair. Surely this is wrong. What I'm advocating here is not a prescription for every PhD. It's a path for only some of us. Having taught well, published articles and papers and books, and created a scholarly identity—the next challenge and source of career fulfillment lies in taking on the job of hiring and mentoring younger scholars, and to start totally new rewarding programs.

8.2 Consolidating the leadership role to grow further, and to stay on top

8.2.1 Stay positive while handling your group/colleagues

At a senior level, juniors would like to make you a co-author in publications under one pretext or another. Resist. Don't encourage co-authorship as their numbers do not matter at your level.

Try to be the best in whatever you are doing. Be persistent. Let no one outwork you.

Look for earnest, good-natured colleagues to act as your sounding board and source of support.

The skulduggery of office politics is unfair, unnecessary, and counterproductive, but you can survive possibly, and maybe even thrive with help from like-minded people there.

8.2.2 Stand behind a junior who is being discriminated against

It is essential for a candidate to consider which of the following performance criteria were considered while rejecting them during a promotion review:
  (1) Your external recommendation letters weren't positive enough.
  (2) Your research is in an emerging or interdisciplinary field, where there typically are fewer researchers, and thus lower citation metrics, even if the underlying work is important and innovative.
  (3) The review committee was not in your favor largely.

(4) Your leadership skills were found lacking.
(5) You spend time on service jobs and on mentorship of younger colleagues, resulting in fewer publications for yourself.

If you are present when you think the candidate was being discriminated against, you should stand behind the candidate and resist them being rejected.

8.2.3 Beware of gaslighting

Gaslighting happens at a workplace when someone tries to change beliefs about who should get the credit or blame for a particular project. That person's restatement of the past about what was to be done specifically, may lead you to question your own memory and to doubt your own interpretation of what happened.

It is natural for people to overestimate their own contributions to a group project a little. But, when someone systematically shifts the conversation about events in their favor and against your interests, it is time to take action. Take notes in meetings to record what specifics were stated. Then you can talk to a trusted mentor or an HR official to share your concerns.

Beware of people who take vindictive glee in petty tyrannies. They are in every workplace—steer clear of them if possible. You may have to work with some people who are very difficult to wok or communicate with. Develop strategies to work with them, and eliminate any frictions.

Sometimes it's not individuals who are the problem but the systems that allow, and in some cases encourage, hostility over cooperation. And systems are hard to change. It is better to try and create a workable situation by re-orienting a bit with a difficult colleague, than hoping things will improve if they leave. In one case, in fact, the difficult colleague decided to stay in the group forever, because a very positive attitude was taken by the senior, to win them over due to their unique capabilities.

8.2.4 Learn to delegate

Learn to delegate rather than shoulder everything yourself. Taking on every challenge isn't leadership. Reach out and seek help.

8.2.5 Leadership during a crisis

When crises land, leaders do one simple yet powerful thing: they seek out and act on the counsel of *other people*. And lots of them. Effective crisis leadership involves gaining perspective in uncertain times by listening to others. Correcting for preconceptions is essential in a crisis, because a crisis is hard to predict and understand in all of its dimensions.

Different leaders navigate a crisis differently and with diverse results. Some flounder at a disruption, like a pandemic. Others emerge more resilient than before. People are studying you closely how you would handle a crisis and will project meaning onto every gesture and offhand comment.

A crisis seldom plays by our established rulebook. Unchecked, a crisis can evolve, expand and engulf in ways we will struggle to handle. Hence, when a crisis hits, you need your leadership to be as bias-free, elastic, deft, and dynamic as the circumstances rapidly unfolding around you.

Keep from listing evergreen and lofty priorities that would only weigh you down. Concentrate on specific actions crucial to set your strategy in motion. When problems pile up, stay focused on what matters most.

A good leader knows that you can't handle a crisis alone. It takes a team to provide the input to forge a vision, create a strategy, and execute that strategy successfully. A good crisis leader knows that when the road gets bumpy, you need a team; but you also need that team to provide or source as many different perspectives on the prevalent situation as possible.

Perspective-taking is a critical skill in crisis management. The more eyes you have on the situation, the less likely it is that you will remain entrenched in your own thinking or anchored to one solution or plan. And the more people you can turn to for counsel as the crisis develops, the easier it will be to shift course and adapt as exigencies dictate. Make sure that you are accessible to anyone who is wanting to share a good idea.

It is important to learn from others, particularly during the middle of a crisis. Learn from others as if you are fresh student of the business. Be open-minded about what opportunities may lie ahead of you.

8.2.6 Remain loyal to your organization

Think always what's best for the organization. That will help you maintain clarity of thinking. As a leader, stand outside someone's individual problem and see holistically to make the best decisions for the organization as a whole.

Crises are inevitable. Even as we leave COVID-19 behind, new clouds may already be gathering. How you can contain the damage or realize the opportunities to emerge more resilient than before? If you don't make judicious use of *all the information* you need to determine *all of the losses and gains* that crises foreshadow, you'll leave yourself and your organization in the dark when you most need to see light at the end of the tunnel.

8.2.7 Share credit, as well as discredit

Never let adversity hold you back. Don't let the fear of failure stop you. People continue in an unsatisfying job because moving to something different daunts them. Rather than not seek a promotion, not stand up to the boss, or not express a controversial opinion, why not think that the worse that could happen to you is that you may lose your job. It won't happen easily, but if it did, then who knows the change might be for the better.

Show appreciation when colleagues respond and let them know how their inputs helped you and prompted a behavioral change.

8.2.8 Be one of them

People learn through personal experience. This eventually becomes explicit knowledge that companies document and use. Different innovations by a company evolve as its people go through different stages of knowledge development, progressively. Spend time on your shop floor to understand the work of your employees and the challenges they face. Sit with them so that you know their difficulties, so that you will not make the mistake of putting forth a non-workable proposition.

Leadership means listening and making people feel included in the environments that we create, as leaders. To learn about your own colleagues, their difficulties at work, listen purposefully. Leave all else in your car, or put it in your pocket, so that you can be present to listen.

When on the shop floor with your staff, use metaphors and stories to help people with different experiences undertake the essence of the business strategy. Honda was a great storyteller and he used a lot of metaphors. For instance, he would often say that a company is like an orchestra, and should be run like it where every person having his or her part in creating new innovations.

One must always respect the views of others in your workplace no matter what they might be. And you must treat everyone around you with respect.

8.2.9 Inspire your team and people around you

As a leader you may have ample confidence, which is there for everyone in your group to see, but your confidence would be seen as a virtue if you also exercise humility, too, while dealing with colleagues. True leaders do have both qualities, simultaneously.

Mentor others to help develop your own leadership skills. Be a confidante to someone who is shy to approach for help, and feels unsafe doing so.

Create a learning culture, rather than a performance culture where the emphasis is solely on results, by acknowledging what you don't know, challenging best practices, and rewarding people who test new ideas even if they don't work. Evidence shows those in learning cultures innovate more and make fewer mistakes.

It is good to know your strengths and weaknesses. Share your skills with others to build relationships. It may be to help a colleague with a presentation. Watching this, the management may consider giving you additional responsibility.

A leader has to push optimism, which implies seeing problems as challenges that are solvable, and to show confidence to do that. Optimists move us forward, as innovators and entrepreneurs. They invest time and effort when they see an opportunity to solve a problem. Pessimists play safe, and let their organization stagnate.

8.2.10 Be innovative and use modern and bold strategies as a leader

Science is based on arriving at a solution that stands the test of time. But for that science recommends to try experiments over and over, to reproduce results. It may be slow, but that is how breakthroughs come.

Leaders have the knack of picking the most promising ideas, which may have weak points, but with some hard work, a leader redesigns them for success. Most innovators succeeded because they were optimists, even though fiercely critical of their own ideas, at times.

Strategy in business depends on forecasting and planning. A leader has to develop critical skills to do lateral thinking to convert resources into useful goods and services, even in times of difficulties, like a global pandemic. Build efficiency during heightened risk by cultivating resilience.

A typical example that describes lateral thinking to convert resources into useful goods and services is a solution discovered and implemented to save the date crops across the warm climate regions of the world from getting afflicted with the red palm weevil. An Israeli company has manufactured a weevil sensor, thousands of which have been drilled into date trees in the UAE, Morocco, and Arab countries. The sensor picks up the vibrations of weevil larvae and sends a warning to an app installed on the farmer's smartphone. As a follow-up of that, a genetically engineered substance is injected into the affected date palm trees and eaten by the weevil. Once ingested, it switches off a vital gene and kills the bug without leaving any chemical trace, as a pesticide would. In fact, the substance, details of which are a trade secret can be genetically engineered into the trees themselves, making them weevil-proof [2].

The wide scope of business and management in the current world has pushed the teaching institutions to go differentially for the functional areas such as marketing, innovation, logistics, communications, entrepreneurship, accounting, and finance. Simultaneously, a leader has to keep an eye on the impact of new technologies such as AI on decision-making, and implement a harmonized approach.

Currently, people are in love with machines (machine learning) and think that machines are going to take over the world. But consider that humans are still at the centre of innovation, because it is humans who have tacit knowledge. And wisdom is just the higher order of tacit knowledge!

8.2.11 Rising on the ladder

One may get very upset, rightfully, on being told 'you are not yet ready for a promotion', even after having completed all the tasks given, and having met all the performance goals. Then why does accomplishing goals not give rewards? Most often, the answer to this is that you have not built or strengthened relationships, which is perhaps a key to promotions.

The senior management has to see that you get along with others, so that a collective performance brings the management more returns. If they find you steamrolling others to achieve goals, then they may consider you a non-cooperating loner. Being willing to change your own mind seems key to getting everyone on the same page.

References

[1] Blow C 2022 My times: career advice from a career in the trenches *New York Times* https://nytimes.com/2022/06/05/opinion/advice-journalism-career.html

[2] Friedman M 2022 The sweet and sticky history of the date *Smithsonian* https://smithsonianmag.com/history/sweet-sticky-history-the-date-180980983/

www.ingramcontent.com/pod-product-compliance
Ingram Content Group UK Ltd.
Pitfield, Milton Keynes, MK11 3LW, UK
UKHW050243150426
5217IPUK00005B/121